Computer-Aided Design in Mechanical Engineering

Computer-Aided Design in Mechanical Engineering

Guest Editor

Rocco Furferi

MDPI

Basel • Beijing • Wuhan • Barcelona • Belgrade • Novi Sad • Cluj • Manchester

Guest Editor
Rocco Furferi
Department of Industrial
Engineering
University of Florence
Florence
Italy

Editorial Office
MDPI AG
Grosspeteranlage 5
4052 Basel, Switzerland

This is a reprint of the Special Issue, published open access by the journal *Applied Sciences* (ISSN 2076-3417), freely accessible at: https://www.mdpi.com/journal/applsci/special_issues/H67NQJ4GAE.

For citation purposes, cite each article independently as indicated on the article page online and as indicated below:

Lastname, A.A.; Lastname, B.B. Article Title. *Journal Name* **Year**, *Volume Number*, Page Range.

ISBN 978-3-7258-8355-4 (Hbk)
ISBN 978-3-7258-8356-1 (PDF)
https://doi.org/10.3390/books978-3-7258-8356-1

Contents

About the Editor vii

Preface ix

Rocco Furferi
Computer-Aided Design in Mechanical Engineering: Insights and Future Directions from a Special Issue
Reprinted from: *Appl. Sci.* **2026**, *16*, 4136, https://doi.org/10.3390/app16094136 1

Rocco Furferi
Integrating Geometric Dimensioning and Tolerancing with Additive Manufacturing: A Perspective
Reprinted from: *Appl. Sci.* **2025**, *15*, 3398, https://doi.org/10.3390/app15063398 12

Jong Hyuk Back, Jeong Bin Bae, Ji Hye Kang and Jung Jin Kim
Comparing Bolt Implementation Methods in Vibration Testing for Accurate Dynamic Behavior Analysis
Reprinted from: *Appl. Sci.* **2025**, *15*, 505, https://doi.org/10.3390/app15020505 22

Mahalingam Nainaragaram Ramasamy, Aleš Slíva, Prasath Govindaraj and Akash Nag
Topology Optimization and Testing of Connecting Rod Based on Static and Dynamic Analyses
Reprinted from: *Appl. Sci.* **2025**, *15*, 2081, https://doi.org/10.3390/app15042081 36

Siyu Zhu, Jie Hu, Jin Qi, Lingyu Wang, Jing Guo, Jin Ma and Guoniu Zhu
Sketch-Guided Topology Optimization with Enhanced Diversity for Innovative Structural Design
Reprinted from: *Appl. Sci.* **2025**, *15*, 2753, https://doi.org/10.3390/app15052753 52

Nikolay Stankov and Aleksandar Ivanov
Automated Design and Parametric Modeling of Excavator Buckets
Reprinted from: *Appl. Sci.* **2025**, *15*, 3389, https://doi.org/10.3390/app15063389 78

Nikolay Stankov and Aleksandar Ivanov
Analytical and Graphical Profiling of Thread-Milling Cutters for Forming Internal Threads
Reprinted from: *Appl. Sci.* **2025**, *15*, 7308, https://doi.org/10.3390/app15137308 111

Sebastian Geyer, Richard Giefing and Christian Hölzl
Development of Customized Algorithms for the Semi-Automatic Generation of Gradient, Conformal Strut-Based Lattice Structures Using Rhino 8 and Grasshopper: Application and Flexural Testing
Reprinted from: *Appl. Sci.* **2025**, *15*, 10364, https://doi.org/10.3390/app151910364 144

Zhenfeng Dai, Zhi Shou, Shuangling Ma and Xiaobo Peng
An Intelligent Process Planning Method for Shaft Parts Based on Multi-Graph Fusion: From Feature Recognition to Process Route Generation
Reprinted from: *Appl. Sci.* **2026**, *16*, 828, https://doi.org/10.3390/app16020828 167

About the Editor

Rocco Furferi

Rocco Furferi is Associate Professor at the Department of Industrial Engineering (DIEF), University of Florence, Italy, conducting research and teaching in the field of industrial engineering design methods, reverse engineering, computer vision, additive manufacturing, and artificial intelligence applications. He received his Ph.D. in Machine Design and Construction from the University of Florence, focusing on artificial vision systems for real-time textile product characterization.

His research interests include AI-driven industrial inspection systems, 2D and 3D computer vision, additive manufacturing, digital manufacturing, reverse engineering, geometric modeling, cultural heritage technologies, biomedical engineering applications, and smart systems for Industry 4.0. He has coordinated and participated in numerous national and international research projects in the fields of textile engineering, heritage science, healthcare technologies, and advanced manufacturing. Notable initiatives include the European FP7 project IMAT for art conservation technologies, the CHANGES PNRR project on cultural heritage innovation, the Tuscany Health Ecosystem initiative on customized medical devices, and several AI-based projects for textile quality control and circular economy.

Prof. Furferi has authored more than 190 scientific publications indexed in Scopus and Web of Science, published in leading journals in mechanical engineering, heritage science, textile engineering, computer vision, and biomedical applications. His work has received international recognition, including Best Paper Awards and acknowledgments for contributions to digital manufacturing, cultural heritage technologies, and 3D printing in healthcare. He also serves as Editorial Board Member and Guest Editor for several international journals, including *Applied Sciences, Heritage Science, Frontiers in Mechanical Engineering,* and *Materials Science in Additive Manufacturing.*

Preface

This Reprint contains a selection of articles discussing the development and future trends of computer-aided design (CAD) technology for mechanical engineering, as well as the incorporation of CAD into new digital manufacturing techniques. The theme revolves around the use of CAD as one of the most relevant supports of the digital design process. The areas covered include parametric modeling, topology optimization, simulation accuracy, additive manufacturing, quality control, and advanced CAD/CAM integration. The book's scope is therefore an update on the latest achievements in research concerning the development of computational design and manufacturing tools and their application in practice. Special emphasis is placed on the idea of the digital thread, where the design process, simulations, manufacturing operations, and validation are connected via digital data environments. The issues addressed in the selected papers concern geometric accuracy, dynamic validation, approaches for generative design, and the application of geometric dimensioning and tolerancing to additive manufacturing systems.

The main purpose of this research is to reveal the new technologies that will form the basis of future design methodology in industrial and, especially, mechanical engineering. The volume shows that modern computer-aided design systems go beyond mere drawings and are becoming intelligent tools for many tasks such as optimization, automation, decision making, and manufacturing integration. At the same time, the influence of artificial intelligence, digital twins, and additive manufacturing in designing innovative and effective solutions is becoming more prominent.

Industries are facing a rise in demand for reduced development time, improved product quality, more efficient structures and material usage, and new validation techniques to verify high-performing mechanical systems. These are the motivating factors behind this Reprint. The spread of generative design through AI, simulation techniques, and additive manufacturing leads to further scientific discussion of these technologies, and their possibilities and obstacles.

The Reprint targets scholars, academics, postgraduate students, and professionals in the disciplines of mechanical engineering, design computation, manufacturing engineering, additive manufacturing, and digital manufacturing processes. It is especially suitable for researchers and practitioners who are keen to learn about advanced techniques in computer-aided design (CAD), topology optimization, simulation-based design, intelligent manufacturing, and artificial intelligence for engineering purposes.

Rocco Furferi
Guest Editor

applied sciences

MDPI

Editorial

Computer-Aided Design in Mechanical Engineering: Insights and Future Directions from a Special Issue

Rocco Furferi

Department of Industrial Engineering, University of Florence, 50134 Florence, Italy; rocco.furferi@unifi.it; Tel.: +39-3204644761

1. Introduction

The discipline of mechanical engineering (ME) relies on digital architecture provided by Computer-Aided Design (CAD). Contemporary CAD systems extend far beyond simple tools for drafting; they are indispensable platforms integrating complex three-dimensional modeling, finite element analysis (FEA), parametric manipulation, and crucial integration points with advanced manufacturing technologies [1]. This radical transformation in the engineering workflow is directly motivated by industry mandates: the necessity to accelerate iterative design cycles, reduce time-to-market intervals, and achieve rigorous optimization of both structural integrity and material efficiency, particularly within safety-critical and high-performance sectors [2]. The benefits of CAD are plain, including significant cost reductions, shorter development cycles, improved product quality, and the ability to rapidly adapt designs to meet changing requirements or constraints. Moreover, CAD serves as an enabler of innovation by providing a platform where cutting-edge technologies can converge. Its integration with virtual and augmented reality opens new horizons for immersive visualization and collaborative design, artificial intelligence enhances decision-making through generative design and automated error detection, and additive manufacturing leverages CAD models directly to create complex, customized structures [3] that were once impossible to fabricate. Together, these advances are expanding the frontiers of what can be achieved in mechanical engineering, fostering a new era of smart, sustainable, and highly efficient product development.

This Special Issue provides a comprehensive overview of the digital thread's current state in ME, presenting novel research that collectively defines the next wave of computational design and manufacturing methodologies. The featured contributions cluster around three interconnected and critical themes. The first theme, related to precision in advanced modeling and Quality Assurance (QA), addresses the fundamental challenge of ensuring geometric accuracy, covering methodologies ranging from proprietary tool profiling techniques to the rigorous establishment of quality frameworks, such as the challenging integration of Geometric Dimensioning and Tolerancing (GD&T) with Additive Manufacturing (AM). The second theme focuses on the revolutionary shift toward sophisticated generative techniques, showcasing advancements that include human-guided and AI-enhanced Topology Optimization (TO) alongside the development of specialized parametric algorithms for creating complex, customized structures. The third theme is related to simulation fidelity and dynamic behavior validation: it explores the absolute necessity of employing rigorous computational and experimental methods to validate dynamic performance, ensuring that digital predictions reliably mirror real-world structural behavior (see Figure 1).

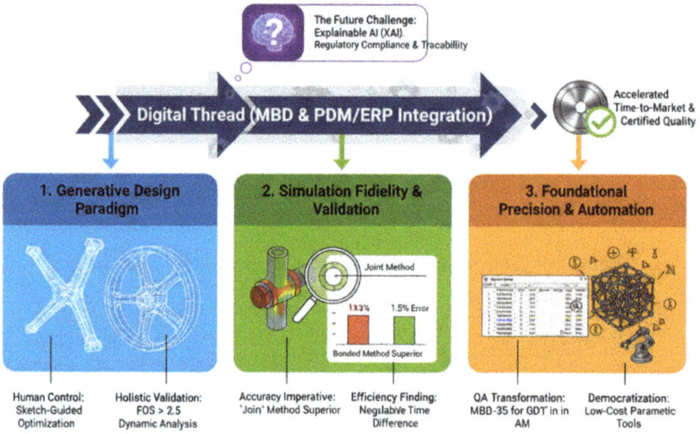

Figure 1. Pillars of the digital threads in papers collected in this Special Issue.

2. Contributions Overview

The articles collected in this Special Issue emphasize an intense industry shift toward customizability, efficiency, and computational rigor in ME design. The range of expertise presented spans from highly technical geometric methodologies to large-scale system integration, as summarized in Table 1.

In [4], a high-precision geometric design required for internal thread-milling cutters is addressed. This work focused on validating the Tangent Circles Method, a graphical profiling technique proposed as an industrially viable and simpler alternative to mathematically complex analytical methods. The comparison demonstrated a "good match" between the profiles derived graphically using CAD software (SolidWorks 2022) and those calculated analytically, confirming deviations were within acceptable manufacturing limits. By effectively democratizing the design process for precision tooling, the method accelerates the design cycle while ensuring the quality of the resulting thread profile. The study confirmed that technological factors like rake angle and relief height also significantly influence the final tool geometry.

The authors in [5] presented a comprehensive methodology for the automated design and parametric modeling of excavator buckets within a CAD environment. The core innovation utilized the configuration abilities of a single 3D model to efficiently manage numerous product variants—ranging broadly in length (600 mm to 2000 mm), features, and type/number of teeth. Quantitative results validated that this parametric approach drastically reduced the overall number of unique models (e.g., from 95 to 35 parts) and corresponding modeling errors by over 50% compared to traditional methods. Crucially, the methodology includes an automated design documentation algorithm that seamlessly integrates the authoritative CAD models with enterprise-wide PDM and ERP systems, ensuring data consistency across the entire digital thread.

The research carried out in [6] introduced a novel multi-solution Topology Optimization (TO) aimed at boosting both the diversity of generated designs and the human designer's control. It proposed integrating human-provided free-hand sketches as quantifiable geometric constraints, separating them into a stroke-based undirected boundary constraint and a closed-shape-based regional constraint. This dual-constraint approach, paired with Fourier mapping-based length-scale control, allows for the flexible manipulation of structural aesthetics without excessively sacrificing mechanical performance. Diversity is enhanced by randomly perturbing internal parameters (like Fourier frequencies) across parallel TO processes. The framework successfully generates diverse, high-performance de-

 https://doi.org/10.3390/app16094136

sign sets for components like wheel spokes, shifting optimization from a singular objective to a controllable design exploration.

Table 1. Key contributions from the papers collected in this Special Issue.

Reference	Primary CAD Theme	Key Contribution/Result	Industrial and Technical Application Context
[4]	Geometric Modeling	Validation of the Tangent Circles Method's high precision compared to analytical profiling.	Internal Thread-Milling Cutters
[5]	Parametric Modeling	Methodology for automated design and parametric configuration modeling using PDM/ERP integration.	Excavator Buckets
[6]	Topology Optimization	Sketch-guided TO framework enhancing solution diversity and human controllability.	Innovative Structural Design (Wheel Spokes)
[7]	Finite Element Analysis	Rigorous validation of TO-designed connecting rod using multi-body dynamics and Goodman fatigue analysis.	Internal Combustion Engine Components
[8]	Simulation Fidelity and Validation	Systematic comparison finding the 'Joint' method is most accurate (7.6 Hz deviation) for dynamic bolt modeling.	Headlamp Vibration Test Jigs
[9]	GD&T and Additive Manufacturing	Perspective on integrating GD&T with AM, focusing on MBD, anisotropy, and novel inspection techniques.	Component Quality Assurance in AM
[10]	Design for Additive Manufacturing	Custom algorithms using Rhino/Grasshopper for semi-automatic generation of gradient, conformal lattice structures.	DfAM Lightweighting
[11]	CAD/CAM Integration and Intelligent Process Planning	Graph-based framework integrating machining feature recognition, process scheme decision-making, and automatic process route generation with high planning accuracy.	Intelligent manufacturing of shaft-type mechanical components

The authors in [7] focused on applying Topology Optimization (TO) to a highly dynamic component—an internal combustion engine connecting rod—with stringent constraints. The objective was to minimize mass and von Mises stress while maintaining a Factor of Safety (FOS) above 2.5, moving beyond simple mass reduction. The optimization process successfully achieved a 5.66% mass reduction from the original design and a 6.25% reduction in maximum von Mises stress. The design was rigorously validated by linking the static TO to a multi-body dynamic analysis (at 1000 rpm) to extract realistic motion loads, followed by a Goodman fatigue life analysis. The final component confirmed long-term durability, with FOS remaining above the critical threshold even at one million cycles.

Reference [8] systematically analyzed the impact of six different FEA bolt connection methods (e.g., bonded, joint, beam) on predicting the dynamic behavior of a structure (a headlamp vibration test jig) against an experimental benchmark. It confirmed that common bonded methods significantly overestimate the natural frequency (544.1 Hz vs. 493.2 Hz experimental), largely due to assumptions of excessive stiffness. The research identified the Joint connection method as the most accurate, achieving a minimal frequency deviation of only 7.6 Hz (a 1.5% error). While bonded methods were the fastest in terms of computational

https://doi.org/10.3390/app16094136

time, the study demonstrated that for high-fidelity dynamic modeling, the minimal extra computing cost of the Joint method yields a massive gain in simulation accuracy.

In the Perspective paper [9], the author outlined the critical challenges and necessary adaptations for integrating established Geometric Dimensioning and Tolerancing (GD&T) principles with Additive Manufacturing (AM) technologies. It highlighted how AM's characteristics—including complex freeform geometries, anisotropic properties, and process-induced variances like warping—break traditional subtractive-based GD&T models. Key advancements proposed include the development of functional datum systems and surface profile tolerancing for non-Euclidean shapes, shifting to area-based surface texture parameters, and the introduction of formalized distinctions between as-built and post-processed tolerances. Fundamentally, the paper advocates for the urgent adoption of Model-Based Definition (MBD) into the Digital Thread to embed all GD&T information directly into the authoritative 3D model.

The authors in [10] addressed the need for accessible tools in Design for Additive Manufacturing (DfAM) by developing custom, semi-automatic Grasshopper algorithms (Rhino 8) for generating sophisticated, strut-based gradient and conformal lattice structures. This low-cost approach challenges the reliance on expensive, proprietary commercial software packages for complex lightweighting solutions. Through three-point bending tests on PA2200 specimens, the research quantified the performance impact of various design parameters. The most significant finding was that Conformity (structures conforming to the boundary geometry) and maximized Strut Diameter positively dominated the resulting load-bearing capacity (force/weight ratio). The work validates the feasibility of democratizing advanced DfAM functionality within customizable parametric modeling environments.

Reference [11] proposes an intelligent process planning framework for shaft-type components that directly bridge CAD geometry and executable machining routes through multi-graph representations. By combining machining feature recognition, process scheme decision-making, and route planning within a unified graph-based architecture, the approach demonstrates how advanced CAD data can be systematically enriched with manufacturing semantics and process constraints. The high reported accuracies in feature recognition and process route generation confirm the potential of graph neural networks to enhance automation, consistency, and decision reliability in CAD/CAM integration, reinforcing the broader trend toward data-driven and intelligent design-to-manufacturing pipelines highlighted throughout this Special Issue.

3. Research Questions and Main Findings

Based on the work of the authors in this Special Issue, a number of research questions and solutions are proposed to stimulate future research in this field.

3.1. *Towards Manufacturing Precision*

In today's industrial landscape, the search for precision in manufacturing is more than a technical goal. Advancements in CAD, digital technologies, automation, and data-driven processes are transforming manufacturing into a discipline where accuracy defines success. Moving towards manufacturing precision means reducing variability, optimizing resources, and delivering products that meet exacting standards while enabling sustainable growth. In this context, the authors of [4] declared that the main research question for high-precision chip-removal processes (such as thread milling) relies in the accurate profiling of the cutting tool as an absolute prerequisite. The traditional analytical methods used to achieve this precision are often mathematically laborious, complex, and computationally intensive, creating a persistent barrier to their widespread industrial adoption. The research presents a solution by validating the Tangent Circles Method, a graphical profiling technique designed for implementation within a standard CAD system, specifically addressing internal thread-milling cutters. The graphical methodology replaces complex equations by defining a series of tangent circles (profiling circles) to an Archimedean spiral (which represents the thread profile in the cross-section). The centers of these circles lie precisely on the specified trajectory of the tool center. The core validation involved comparing the geometric parameters that were graphically derived with the results obtained from complex analytical equations. The comparison demonstrated a "good match," confirming that the resulting profile deviations were within acceptable manufacturing limits. This outcome validates the Tangent Circles Method as being both mathematically sound and industrially viable.

3.2. Enhancing Modeling Efficiency Through Parametric Automation and PDM/ERP Integration

The engineering challenges in managing extensive product lines are characterized by high-variability results in model proliferation, redundant efforts, and a significant increase in potential design errors [12]. To deal with the relevant research question, a methodology is introduced that fundamentally addresses this issue by leveraging parametric modeling via configurations within a CAD system such as SolidWorks 2022 [5].

This approach is exemplified by its application to excavator buckets, which range broadly from Class A to Class M with lengths spanning 600 mm to 2000 mm. Components, including the Base Part and Upper Part, are designed parametrically, allowing diverse product versions to be managed within a single 3D model. The quantifiable efficiency gains are substantial. Comparative data confirms a reduction in the total number of unique models required, shrinking from 95 parts in the traditional approach to only 35 parts, resulting in average time-savings of over 50% for modeling (e.g., 65 h reduced to 34 h for parts). Crucially, this reduction in manual effort yields a proportional reduction in observed modeling errors.

The proposed methodology formalizes the link between CAD and enterprise-wide systems. It explicitly incorporates the integration of CAD models with Product Data Management (PDM) systems (such as SolidWorks PDM 2022) and Enterprise Resource Planning (ERP) systems (such as RDS ERP). While this initial setup requires a high upfront investment in specialized knowledge, the resultant long-term benefit is a non-linear scaling of design efficiency. PDM/ERP integration transforms the engineering process from a sequence of compartmentalized tasks into an integrated system where manufacturing and logistics (e.g., Supplies and Production) instantly access accurate, verified design data. This approach establishes the CAD model as the authoritative data source for the entire enterprise, reinforcing the Digital Thread concept and delivering systemic efficiency through centralized data management and improved traceability, which is vital for accelerating time-to-market.

3.3. Next Generation Topology Optimization: Human Control and Dynamic Evaluation

Topology Optimization (TO) is widely adopted to generate high-performance structures, typically aiming to minimize compliance under volume constraints [13]. However, traditional TO algorithms suffer from two main limitations: limited human controllability over design features and insufficient solution diversity, often necessitating burdensome manual redesign iterations. To deal with this issue, a novel sketch-guided, multi-solution TO framework is proposed in [7] to overcome these limitations by intentionally integrating human aesthetic input and computational randomness. This framework operates by codifying intuitive human aesthetic and stylistic preferences (expressed via sketch) as soft, quantifiable geometric constraints that are designed to dynamically compete with the primary engineering objective (performance/compliance minimization).

 https://doi.org/10.3390/app16094136

The framework uses dual control strategies, integrated with Fourier mapping-based length-scale control: a Stroke-based undirected structural boundary constraint ensures the structural boundaries align with sketched strokes. This constraint preserves design freedom in adjacent regions and is vital for promoting design diversity and helping the algorithm escape localized optima, although it may result in a slight increase in compliance. A closed-shape-based regional constraint drives material uniformity within a closed sketch contour. This constraint effectively stabilizes structural features and approximates a hard constraint when given a large weight coefficient, although its use tends to reduce solution diversity. Diversity is significantly enhanced by running multiple parallel TO processes where two key parameters are randomly perturbed: the Fourier mapping frequencies and the load force magnitudes. This strategy manages the adversarial relationship between aesthetic control and the optimization objective to produce a set of diverse, high-performing solutions. The result is that the designer shifts from passively accepting one singular optimal solution to actively exploring a Pareto front defined by both performance metrics and human-defined aesthetic criteria.

Reference [7] explores the potential of next generation TO with particular focus on its application in highly dynamic components within internal combustion engines, such as connecting rods, mandates a stringent validation process that extends far beyond standard static stress analysis. A rigorous workflow is presented that systematically integrates mass/stress minimization (using the Solid Isotropic Material with Penalization or SIMP method in ANSYS Discovery 2024 R2) with an extensive post-optimization validation incorporating rigid body dynamics, transient analysis, and fatigue life analysis. The optimization successfully targeted mass reduction and maximum von Mises stress minimization, while simultaneously adhering to the critical constraint of maintaining a reliable factor of safety.

This research strongly emphasizes that for highly dynamic systems, successful TO is insufficient without rigorous, multi-physical validation that accounts for static compression, complex dynamic/inertial loads, and longevity. Critically, the study found that excessive mass reduction leads to noticeable engine imbalance and increased vibration due to a disturbance in the inertial balancing ratio between the connecting rod mass and the crankshaft counterweight. This reinforces that optimization in dynamic systems must be a holistic process constrained by kinematic and inertial requirements, not merely compliance and volume.

A complementary research direction emerges at the interface between CAD and manufacturing planning [11], where graph-based artificial intelligence methods extend design intelligence beyond geometry toward automated decision-making across the downstream stages of production. In this context, CAD models are no longer treated as static repositories of shape information, but as structured data sources enriched with manufacturing semantics, process constraints, and operational knowledge. By leveraging graph representations and learning-based reasoning, these approaches enable the automatic selection of machining schemes and the generation of consistent process routes directly from design data, reinforcing the role of CAD as the central driver of an integrated, intelligent design-to-manufacturing pipeline.

3.4. Integrating Quality Control in Design for Additive Manufacturing: The Challenge of GD&T Standardization

Additive Manufacturing (AM) enables the creation of highly innovative and complex geometries, such as lattice structures, gyroids, and inaccessible internal channels [14]. Simultaneously, AM processes introduce undesired effects to be controlled, such as anisotropic material properties, thermal distortion, warping, and shrinkage [15]. These factors challenge the core tenets of Geometric Dimensioning and Tolerancing (GD&T), which was historically developed and standardized for subtractive processes. Traditional GD&T stan-

 https://doi.org/10.3390/app16094136

dards struggle with non-Euclidean surfaces, where defining datum systems for freeform shapes is non-trivial; anisotropy, where layer-by-layer fabrication causes significantly rougher surfaces in the vertical/build direction; and inspection limitations, where internal features are inaccessible to traditional Coordinate Measuring Machines (CMMs).

This situation is not merely a metrology issue; it represents a fundamental challenge to the industrial scalability and certification of AM parts. Without robust, unambiguous, and standardized tolerancing, the innovative designs generated by DfAM tools cannot move reliably into regulated mass production [9], especially when integrated into complex assemblies alongside subtractive manufactured components. To address this critical bottleneck, the perspective highlights several necessary adaptations in GD&T practice:

- A transition toward functional datum systems and surface profile tolerancing for freeform geometries, validated using non-contact metrology like point cloud analysis or mesh comparison.
- The evolution of surface roughness standards, shifting focus from traditional line-based parameters (*Ra*, *Rz*) to area-based parameters (*Sa*, *Sq*), which are far better-suited to characterizing complex, anisotropic AM surfaces.
- The formal integration of manufacturing fidelity requirements by distinguishing between as-built tolerances (raw AM surface finish) and post-processed tolerances (tighter limits achieved through secondary operations).

The solution requires fully integrating Model-Based Definition (MBD) into the Digital Thread. MBD embeds all GD&T information, material specifications, and necessary annotations directly into the 3D model, eliminating the reliance on ambiguous 2D drawings. This integration facilitates automated tolerance checks during build simulation and enables advanced inspection techniques to verify the component directly against the CAD source, ultimately unlocking the full potential of complex generative designs for industrial scale.

3.5. Customization and Conformity in Design for Additive Manufacturing

The capacity to design complex, functionally graded lattice structures is traditionally restricted by reliance on highly specialized, expensive commercial software (e.g., nTopology nTop or Altair Inspire Personal Edition), creating significant financial and accessibility barriers for research and small firms. As already mentioned, a methodology is developed in [10] for generating customized, semi-automatic Grasshopper algorithms (a Rhino 8 plugin) to generate gradient, conformal, strut-based lattice structures, offering a low-cost, customizable alternative. The custom algorithm leverages native Grasshopper functions combined with open-source plugins (Crystallon V2.0, Dendro V0.9.0, IntraLattice V07.06, Pufferfish V2.3) to control complex parameters, including unit cell topology, conformity (conforming or non-conforming to surface geometry), and gradients in cell size or strut diameter. The functionality was subjected to quantitative experimental validation using 3-point bending tests on PA2200 arch specimens, focusing on the resulting force/weight ratio. This outcome confirms the potential of leveraging accessible parametric CAD environments with open-source extensions to achieve complex DfAM functionality previously restricted to proprietary commercial packages. This represents a significant pathway toward democratizing advanced lightweighting and material performance optimization techniques by lowering the financial and learning curve barriers.

4. Discussion

The findings of the seven papers published in this Special Issue collectively demonstrate the existence of a robust, though continuously maturing, digital engineering ecosystem in ME:

https://doi.org/10.3390/app16094136

- Validation Requires Multi-Physical Integration: The development of high-performance components demands that validation protocols transcend disciplinary silos, linking geometry (TO) with structural mechanics (stress), dynamics (kinematics/vibration), and manufacturing feasibility.
- The Trend of Accessible Customization: The industry exhibits a shift away from single dependence on monolithic software. Evidence suggests that custom, bespoke algorithms built on accessible platforms (such as Rhino/Grasshopper) can successfully deliver highly specialized DfAM results, such as conformal and gradient lattice structures, which were previously restricted to expensive commercial packages.
- Data Centrality in Quality and Efficiency: Both efficiency gains and rigorous quality assurance fundamentally rely on robust, centralized data management systems (PDM/ERP) and the establishment of the CAD model, via MBD, as the single authoritative source of product data throughout its lifecycle.

The synthesized research points toward a convergence of the technologies that will define the next decade of CAD/ME practice, marked by the increasing influence of cognitive systems and the formalization of digital data pipelines. Therefore, the following areas of investigations will be the subject scientific and industrial research in the near future.

4.1. The Rise in Artificial Intelligence for Generative Design

The integration of Artificial Intelligence is one of the most frequently cited directions for CAD evolution [16], promising to automate routine tasks, suggest methods of material optimization, and provide predictive insights by learning from the historical data and simulation outcomes. Generative Design is concurrently expanding its reach, moving beyond simplistic mechanical constraints to integrate multi-disciplinary considerations, including thermal, electrical, and kinematic factors [17].

However, the primary long-term challenge for the adoption of AI in ME industries (e.g., automotive and aerospace) is not technical capacity but regulatory compliance and traceability. Current AI solutions, particularly generative models (LLMs), frequently operate as "black boxes," making it difficult to trace their engineering rationale. This lack of transparency undermines the traceability required by standards such as the Machinery Directive 2006/42/EC [18] and ISO 12100.4 [19].

The critical future research direction for AI in ME must focus on developing Explainable AI (XAI) models for Topology Optimization and GD&T that can provide verifiable, traceable design rationale [20], and establishing AI-based compliance checking mechanisms to ensure automatically generated designs satisfy safety and regulatory standards before physical prototypes are manufactured.

4.2. The Evolution of Digital Thread and Digital Twin

The realization of Digital Thread is an essential condition for achieving industrial scaling in AM [21]. The successful implementation of the digital thread hinges on fully adopting the Model-Based Definition (MBD), which ensures that GD&T annotations and other critical parameters are embedded directly into the authoritative 3D model, thereby minimizing data loss and misinterpretation across the chain (CAD, Simulation, and CAM).

High-fidelity parametric models and accurate dynamic simulations serve as the digital foundation for the Digital Twin concept. The logical progression is to integrate these virtual models with real-time operational data, allowing engineers to track wear, simulate predictive maintenance, and refine subsequent designs based on true operating conditions. Future research efforts should prioritize the integration of MBD data [22] and parametric modeling systems with real-time operational data streams, especially for complex machinery like excavator buckets, to enable continuous, data-driven design optimization.

 https://doi.org/10.3390/app16094136

4.3. Gaps in Standardization and Validation Processes

The points identified across the collected research necessitate targeted efforts in the development of standards. Dealing with AM metrology, as well as the complexities of AM geometries, particularly internal features and freeform surfaces, demand further standardization of non-contact inspection techniques (e.g., Computed Tomography (CT) scanning and optical 3D scanning) to ensure reliable quality verification [9].

When referring to dynamic modeling Standards, it is evident that the quantitative findings regarding simulation fidelity should inform industry guidelines on the preferred bolt modeling methods for highly dynamic or vibration-sensitive structures. These standards should explicitly balance the need for high accuracy (joint method) against the computational and complexity overhead (bonded method) across different design phases.

Finally, with reference to multi-material tolerancing, the current GD&T frameworks are insufficient for handling the dimensional complexity of multi-material or functionally graded AM parts, requiring new standards to establish consistent and unambiguous tolerance values. All these aspects should be considered for future research in the field of CAD for ME.

5. Conclusions

This Special Issue highlights the current trend in Computer-Aided Design for Mechanical Engineering, characterized by a search for efficiency and accuracy. The contributions collectively demonstrate significant advances:

- Efficiency Through Parametric Systems [23]: The proven feasibility of employing parametric modeling integrated with PDM/ERP systems substantially reduces design time and error for product lines with high variability, institutionalizing the digital thread.
- Precision: The validation of graphical, intuitive CAD techniques, such as the Tangent Circles Method, for traditionally analytical precision tool profiling, enhances the accessibility of complex engineering geometry in standard industrial practices [4].
- Generative Intelligence: The development of novel TO frameworks successfully integrates human stylistic preferences (sketches) as functional constraints that simultaneously drive the generation of diverse, high-performance design sets, marking a pivotal step in human–machine collaboration [24].
- Rigorous Validation: The establishment of comprehensive protocols for generative designs mandates analysis across static stress, multi-body dynamics, and fatigue life for dynamic components [25].
- DfAM Maturation: The ability to generate complex, functionally graded lattice structures using customized, low-cost parametric tools challenges the dominance of high-cost commercial software and expands access to advanced DfAM [7,8].
- Quality Assurance: The urgent need for GD&T to adapt via MBD, area-based parameters, and non-contact inspection technologies is highlighted as essential for reliably managing the complexities inherent in AM geometries [9].

The future path of CAD in Mechanical Engineering is moving toward fully cognitive, AI-driven digital methodologies and tools, where design, manufacturing, and validation are seamlessly and intrinsically linked. This forthcoming generation of tools will rely on the foundations established here: robust parametric modeling, highly accurate simulation fidelity, and formalized digital communication. Future research efforts, particularly those aimed at resolving the fundamental regulatory and traceability challenges posed by AI, the full realization of the Digital Twin concept, and the necessary standardization of GD&T frameworks for Additive Manufacturing, will ultimately define the industrial competitiveness of the next generation of mechanical components.

https://doi.org/10.3390/app16094136

Funding: This research received no external funding.

Conflicts of Interest: The author declares no conflicts of interest.

References

1. Reddy, E.J.; Reddy, K.S.; Reddy, P.N. Recent advancements in knowledge-based parametric modeling of mechanical components. In *International Conference on Mechanical and Energy Technologies*; Springer Nature: Singapore, 2023; pp. 401–411.
2. Alsubaih, A.A.S.; Sepehrnoori, K.; Delshad, M.; Alsaedi, A. A Comprehensive Review of Well Integrity Challenges and Digital Twin Applications Across Conventional, Unconventional, and Storage Wells. *Energies* **2025**, *18*, 4757. [CrossRef]
3. Buonamici, F.; Furferi, R.; Governi, L.; Lazzeri, S.; McGreevy, K.S.; Servi, M.; Talanti, E.; Uccheddu, F.; Volpe, Y. A CAD-based Procedure for Designing 3D Printable Arm-Wrist-Hand Cast. *Comput.-Aided Des. Appl.* **2019**, *16*, 25–34.
4. Stankov, N.; Ivanov, A. Analytical and Graphical Profiling of Thread-Milling Cutters for Forming Internal Threads. *Appl. Sci.* **2025**, *15*, 7308. [CrossRef]
5. Stankov, N.; Ivanov, A. Automated Design and Parametric Modeling of Excavator Buckets. *Appl. Sci.* **2025**, *15*, 3389. [CrossRef]
6. Zhu, S.; Hu, J.; Qi, J.; Wang, L.; Guo, J.; Ma, J.; Zhu, G. Sketch-Guided Topology Optimization with Enhanced Diversity for Innovative Structural Design. *Appl. Sci.* **2025**, *15*, 2753. [CrossRef]
7. Nainaragaram Ramasamy, M.; Slíva, A.; Govindaraj, P.; Nag, A. Topology Optimization and Testing of Connecting Rod Based on Static and Dynamic Analyses. *Appl. Sci.* **2025**, *15*, 2081. [CrossRef]
8. Back, J.H.; Bae, J.B.; Kang, J.H.; Kim, J.J. Comparing Bolt Implementation Methods in Vibration Testing for Accurate Dynamic Behavior Analysis. *Appl. Sci.* **2025**, *15*, 505. [CrossRef]
9. Furferi, R. Integrating Geometric Dimensioning and Tolerancing with Additive Manufacturing: A Perspective. *Appl. Sci.* **2025**, *15*, 3398. [CrossRef]
10. Geyer, S.; Giefing, R.; Hölzl, C. Development of Customized Algorithms for the Semi-Automatic Generation of Gradient, Conformal Strut-Based Lattice Structures Using Rhino 8 and Grasshopper: Application and Flexural Testing. *Appl. Sci.* **2025**, *15*, 10364. [CrossRef]
11. Dai, Z.; Shou, Z.; Ma, S.; Peng, X. An Intelligent Process Planning Method for Shaft Parts Based on Multi-Graph Fusion: From Feature Recognition to Process Route Generation. *Appl. Sci.* **2026**, *16*, 828. [CrossRef]
12. Metzger, A.; Pohl, K. Software product line engineering and variability management: Achievements and challenges. In Proceedings of the Future of Software Engineering Proceedings, Hyderabad, India, 31 May–7 June 2014; pp. 70–84.
13. Meli, E.; Furferi, R.; Rind, A.; Ridolfi, A.; Volpe, Y.; Buonamici, F. A general framework for designing 3D impellers using topology optimization and additive manufacturing. *IEEE Access* **2020**, *8*, 60259–60269. [CrossRef]
14. Uccheddu, F.; Carfagni, M.; Governi, L.; Furferi, R.; Volpe, Y.; Nocerino, E. 3D printing of cardiac structures from medical images: An overview of methods and interactive tools. *Int. J. Interact. Des. Manuf.* **2018**, *12*, 597–609. [CrossRef]
15. Wang, Y.; Blache, R.; Xu, X. Selection of additive manufacturing processes. *Rapid Prototyp. J.* **2017**, *23*, 434–447. [CrossRef]
16. Enahoro, M.O.; Ekiugbo, A. Evolution of CAD in Mechanical Engineering. *J. Mater. Eng. Struct. Comput.* **2015**, *4*, 122–137.
17. Buonamici, F.; Carfagni, M.; Furferi, R.; Volpe, Y.; Governi, L. Generative design: An explorative study. *Comput.-Aided Des. Appl.* **2021**, *18*, 144–155. [CrossRef]
18. Directive 2006/42/EC17, May 2006, Official Journal of the European Union. Available online: https://eur-lex.europa.eu/eli/dir/2006/42/oj/eng (accessed on 20 April 2026).
19. *UNI EN ISO 12100:2010*; Safety of Machinery—General Principles for Design—Risk Assessment and Risk Reduction. ISO: Geneva, Switzerland, 2010.
20. Trivedi, C.; Bhattacharya, P.; Prasad, V.K.; Patel, V.; Singh, A.; Tanwar, S.; Sharma, R.; Aluvala, S.; Pau, G.; Sharma, G. Explainable AI for industry 5.0: Vision, architecture, and potential directions. *IEEE Open J. Ind. Appl.* **2024**, *5*, 177–208. [CrossRef]
21. Zhang, Q.; Liu, J.; Chen, X. A literature review of the digital thread: Definition, key technologies, and applications. *Systems* **2024**, *12*, 70. [CrossRef]
22. Bocciarelli, P.; D'Ambrogio, A.; Falcone, A.; Garro, A.; Giglio, A. A model-driven approach to enable the simulation of complex systems on distributed architectures. *Simulation* **2019**, *95*, 1185–1211. [CrossRef]
23. Touloupaki, E.; Theodosiou, T. Performance Simulation Integrated in Parametric 3D Modeling as a Method for Early Stage Design Optimization—A Review. *Energies* **2017**, *10*, 637. [CrossRef]

 https://doi.org/10.3390/app16094136

24. Charalampous, P. A Computational Sketch-Based Approach Towards Optimal Product Design Solutions. *Appl. Sci.* **2025**, *15*, 2413. [CrossRef]
25. Jiménez, C.A.; Hernández Gálvez, G.; Dorrego Portela, J.R.; Verde Añorve, A.; Ibáñez Duharte, G.; Pantoja Enríquez, J.; Lastres Danguillecourt, O.; Perea-Moreno, A.-J.; Muñoz-Rodriguez, D.; Hernandez-Escobedo, Q. Sustainable Analysis of Wind Turbine Blade Fatigue: Simplified Method for Dynamic Load Measurement and Life Estimation. *Sustainability* **2025**, *17*, 7615. [CrossRef]

https://doi.org/10.3390/app16094136

applied sciences

MDPI

Perspective

Integrating Geometric Dimensioning and Tolerancing with Additive Manufacturing: A Perspective

Rocco Furferi

Department of Industrial Engineering, University of Florence, 50139 Firenze, Italy; rocco.furferi@unifi.it; Tel.: +39-3204644761

Abstract: Geometric Dimensioning and Tolerancing (GD&T) are among the basic concepts of functional fitness and quality assurance in modern manufacturing. The historical development of GD&T took place primarily in the ambit of subtractive manufacturing; the advent of Additive Manufacturing (AM) now presents novel challenges due to the complexity of geometries, material variability, and process-induced variances. The present Perspective Paper briefly hints at key challenges for the future of GD&T in AM, with an eye to the necessary adaptation of tolerancing principles to AM-specific geometries, integration of Model-Based Definition (MBD) in digital threads, and development of new standards for surface texture and tolerance stack-up. New inspection techniques are also highlighted for the AM parts, which would become more prominent. This study underscores the need for continued research and collaboration to develop comprehensive GD&T frameworks tailored to AM, ensuring its industrial scalability and interoperability with traditional manufacturing systems.

Keywords: geometric dimensioning and tolerancing; ISO; additive manufacturing; model-based definition; surface texture features

1. Introduction

As widely recognized, Geometric Dimensioning and Tolerancing (GD&T) is a method for specifying and conveying engineering tolerances on engineering drawings and computer-generated 3D models, using a symbolic language that explains the nominal geometry of a physical object and the acceptable variations from it [1]. The nominal (theoretically perfect) geometry of parts and assemblies, the permitted variation in the size, form, orientation, and location of individual features, and how features may differ from one another so that a component is deemed adequate for its intended use are all defined by GD&T. This framework addresses traditional subtractive manufacturing techniques like machining, turning, and milling, and establishes geometric and dimensional criteria for designed components [2,3]. These processes allow the production of either simple parts with prismatic or revolving features or more complex geometries (especially when CNC machine centers are adopted); in any case, such parts are characterized by a well-defined topology and, depending on the process itself, by specific and usually consistently annotatable tolerance values. Quite the reverse, Additive Manufacturing (AM) enables the creation of complex geometries and freeform shapes that challenge traditional tolerancing approaches [4]. Mainly due to the special features of additive processes, including layer-by-layer fabrication, the manufacturing of complex geometries, and material variability, the adoption of GD&T in AM is expected to be significantly developed in the next years. Standardized techniques to properly specify and communicate dimensions and geometric tolerances of additive manufactured parts will be required to deal with issues like the implementation of topology-optimized designs, surface roughness, and the presence of anisotropic features on a macroscopic scale (e.g., resulting from layer-by-layer manufacturing) and on a microscopic scale (e.g., due to the thermal history in the AM process). In particular, the main challenge is to ensure that AM components fulfill functional requirements while aiding efficient production and quality control, especially when integrated into machines composed of both subtractive manufactured and AM components. To achieve dimensional compliance with the original CAD model, some approaches are used to manage geometric, dimensional, or shape errors in AM. For instance, stair-stepping, a rough surface caused by AM's layer-by-layer bottom-up method, can potentially result in a geometrical error. Along with component shrinkage, these geometrical errors are also influenced by material qualities, process parameters, support structures, and surface approximation errors brought on by slicing techniques. As a consequence, the design phase needs to take into account the expected actual geometry of the manufactured parts, planning a set of actions to ensure its compliance with the originally designed digital part. Accordingly, the geometrical mismatch between the printed items and the CAD and toolpath models is a significant problem to be confronted with when AM is involved as a primary method for building a part [5]. Because of the alterations made during the slicing step to build a toolpath model that depends on different slicing settings, the printed part is not the same as the original CAD model. The goal is to reduce the deviations and/or compensate them based on the process conditions [6]. However, to function as a comprehensive geometric quality sensing and control strategy, the integration of GD&T and AM requires considerable development. According to recent literature [6–10], the key perspectives for integrating GD&T and AM can be categorized into the following issues:

- Adaptation of GD&T principles to AM-specific geometries
- Surface texture and tolerance stack-up
- Digital thread and model-based definition [11]
- The advent of Artificial Intelligence in GD&T
- Standardization of inspection for AM
- GD&T adaptation to different technologies
- GD&T in AM assembly parts

The present Perspective Paper aims to briefly hint at key challenges for the future of GD&T in AM, with an eye to the necessary adaptation of tolerancing principles to AM-specific geometries, integration of Model-Based Definition (MBD) in digital threads, and development of new standards for surface texture and tolerance stack-up. This will act as a basis for stimulating a discussion within the scientific community on this fascinating topic that will be prominent shortly in manufacturing and industrial fields.

Figure 1 shows a conceptual schematic illustrating the adaptation of GD&T principles to Additive Manufacturing.

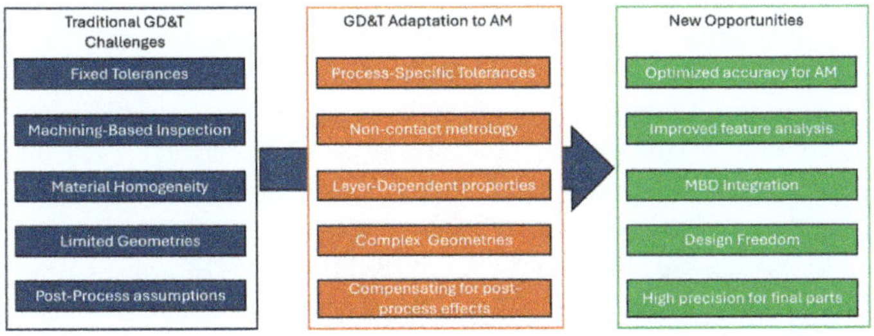

Figure 1. Conceptual schematic illustrating the adaptation of GD&T principles to AM.

2. Adaptation of GD&T Principles to AM-Specific Geometries

As mentioned above, differently from subtractive manufacturing, AM introduces some design complexities such as organic geometries, the presence of internal features, and, possibly, lattice structures. Dealing with non-Euclidean surfaces, these are difficult to dimension and tolerate using conventional GD&T standards. By way of example, lattice structures or gyroids used in lightweight aerospace components may not have clearly defined planes, axes, or cylinders for referencing traditional datum systems. For what concerns internal features, such as channels, cavities, and voids, the main issue is related to their inaccessibility for measurement using traditional inspection tools like coordinate measuring machines (CMMs); as a consequence, it is not trivial to determine the actual dimensions of such features, despite the fact that this challenge is shared with the field of precision casting, where several techniques can also be adapted to AM. Finally, the layer-by-layer process in AM results in anisotropic material properties and dimensional variation due to factors like residual stress, heat distortion, or powder bed characteristics. This makes it difficult to define a set of geometrical tolerances on the part.

To address these challenges, GD&T is rapidly evolving to encompass functional datum systems that can reference freeform surfaces instead of relying on idealized planes or cylinders [12,13]. Therefore, future GD&T practices will integrate surface profile tolerancing for freeform surfaces using methods such as point cloud analysis or mesh comparison to verify the manufactured geometry against the design intent. Not by chance, ISO 21920-1:2021 [14] and ASME Y14.46 [15] (under development for AM-specific GD&T) define a set of parameters for surface texture, including those suitable for irregular shapes produced by AM. AM has been used by businesses like Rolls-Royce and GE Aviation for manufacturing engine parts with complex geometries, such as fuel nozzles. Tolerances for freeform surfaces and functional interfaces need to be defined for these components using GD&T modifications. Another example is provided by freeform designs customized to each patient's anatomy, which are used in patient-specific implants, such as those made with AM for orthopedics. Instead of using traditional dimensions, GD&T standards should be adopted to include tolerances for fitness and function.

3. Surface Texture and Tolerance Stack-Up

Since AM is mostly layer-by-layer based, tolerance stack-up and surface texture requirements are different from those adopted in conventional manufacturing. Layer thickness, material deposition technique, powder size, and post-processing methods are some of the parameters that affect both the final shape of the part (which deviates from the designed part and the traditional machined surfaces) and the surface finish in AM parts. These variations provide difficulties for conventional GD&T procedures and ISO tolerancing standards.

In terms of surface texture, it should be noticed that the layering process causes AM parts to generally have anisotropic surface roughness, with the vertical (build) direction frequently having rougher surfaces than the horizontal. As a consequence, standard roughness parameters such as Ra (Arithmetic Average Roughness), Rz (Maximum Height), or Sa (Surface Area Roughness) may not fully define the complexity of AM surface profiles [16]. Furthermore, AM may use support structures to stabilize overhanging features [17], leaving residual marks or artifacts that affect the surface finish. The immediate consequence is that designers should be aware of the difference between specific tolerances for post-processed areas and as-built surfaces. As seen by ISO/ASTM 52920:2023 standard [18], which establishes rules for test artifacts in AM to assess surface roughness, standards are progressively concentrating on specifying surface finish requirements, especially for AM procedures. Moreover, several authors [19–21] are proposing the use of area-based parameters (Sa, Sq) to better characterize roughness in AM parts compared to the traditional line-based parameters (e.g., Ra). In addition, inspection methods based on Reverse Engineering [22,23] methods, such as optical 3D scanning, light interferometry, and CT, can be used for assessing complex AM surfaces, thus allowing GD&T standards to enable precise evaluation of surface profiles.

Regarding tolerance stack-up, dimensional variability in individual parts brought on by AM process features (such as warping or shrinking) might result in unforeseen tolerance stack-up problems in assemblies, including AM components. For instance, when tolerances are not properly maintained across many AM components, parts with tight assembly clearances (e.g., turbine blades) cannot operate under design conditions. To deal with this issue, GD&T practices will increasingly include provisions for as-built tolerances, which reflect the raw AM surface finish directly from the 3D printer, and post-processed tolerances that will ensure tighter limits achieved through secondary processes such as machining, polishing, or chemical finishing [24]. Additionally, process-induced variations like warping and thermal distortions will be incorporated into tolerance stack-up models for AM parts, and sophisticated simulation techniques like digital twins and finite element analysis (FEA) will aid in the prediction of tolerance accumulations in assemblies.

Unfortunately, since existing GD&T standards primarily focus on single-material components, it is still difficult to obtain consistent tolerance values for multi-material or graded-material AM parts. As a result, future studies are anticipated to focus on this area [25].

4. Digital Thread and MBD Integration

A recent improvement in Additive Manufacturing is the combination of Model-Based Definition (MBD) with GD&T within the digital thread ecosystem. By ensuring that tolerance criteria are efficiently transferred from design to both manufacture and inspection, this integration is expected to improve the AM workflow's accuracy and efficiency. It is expected in the near future that GD&T information, material specifications, and other annotations will be embedded directly into a 3D CAD model, eliminating the reliance on traditional 2D drawings. Even though we live in an increasingly digital world, this central step is not easy to implement immediately. Many companies, especially SMEs, still work using 2D drawings for manufacturing. This requires the designer to indicate the geometric and dimensional tolerances on this drawing, rendering the analysis of AM processes more complex. By empowering MBD, even for traditional subtractive processes, it will be easier to consider the precise definition of tolerances on simple geometries but also on freeform surfaces and lattice structures. ISO standards are aware of this opportunity, as demonstrated by ISO 16792:2021 [26]—Digital Product Definition Data Practices, which specifies requirements for the preparation, revision, and presentation of digital product definition data, and supports two methods of application: 3D model only and 3D model with 2D drawing in digital format. Under these premises, organizations are turning to Additive Manufacturing (AM) as a solution to supply chain and manufacturing restrictions in today's more complicated production and sustainability contexts. However, companies must oversee intricate, interconnected, and data-driven events to deploy AM on an industrial scale. The secret to enhancing AM capability is the digital thread, a single, seamless strand of data that runs from the original design to the completed part.

Digital thread's primary benefit is that it allows designers to directly implement GD&T into 3D models, ensuring their tolerance objectives are maintained throughout the manufacturing and inspection processes. When specifying tolerances on complex surfaces and interior aspects that are difficult to accurately depict in 2D drawings, this is especially helpful. Additionally, automatic tolerance checks are made easier during the process simulation and build preparation phases by incorporating GD&T within the

MBD framework. In fact, automatic GD&T assignment and validation for AM-specific geometries is now supported by software programs like Siemens NX version 2306 (Siemens AG, Munich, Germany), PTC Creo (PTC Inc., Boston, MA, USA) and Autodesk Fusion 360 (Autodesk Inc. San Francisco, CA, USA). Lastly, because inspection tools may extract GD&T information directly from CAD models, GD&T annotations in the digital thread allow automated inspection utilizing sophisticated measuring systems like 3D scanners and CT scans, minimizing errors in manual interpretation. However, even though MBD is supported by several GD&T standards (such as ISO 1101 [27]), further harmonization is needed to handle AM's particular requirements, like layer-specific annotations and freeform tolerancing. Furthermore, data loss or incorrect interpretation of GD&T annotations in AM processes may result from a lack of connectivity between design, simulation, and manufacturing software. Fortunately, this issue was addressed in the past by leveraging the concept of Constructive Technological Entity that aims to embed the same object pieces of geometrical and technological information, targeting a better data transfer between different types of software (e.g. CAD like Siemens NX, version 2306 and CAM like Siemens NX CAM version 2024). As a final remark, it is possible to state that Artificial Intelligence-based methods will also assume their role in this process by leveraging the integration of GD&T and MBD by automating tolerance assignment based on part functionality and AM process capabilities [28]. By analyzing data from previous builds, machine learning models can anticipate deviations and adjust process parameters, leading to improved accuracy and reduced need for post-processing. This data-driven approach enhances the reliability and efficiency of AM processes [29].

MBD is used by firms such as Lockheed Martin, Boeing, and Baker Hughes to include GD&T into AM component designs, guaranteeing adherence to stringent aerospace tolerances throughout manufacturing and inspection. For AM parts, automakers such as Ford and BMW use MBD and GD&T integration to speed up prototyping and guarantee dimensional correctness during assembly.

5. The Advent of Artificial Intelligence in Geometric Dimensioning and Tolerancing (GD&T)

The rapid development of Artificial Intelligence may be considered among the hottest topics in research in countless industry sectors today. GD&T makes no exception. Recent works deal with the implementation of AI-based methods, mostly making use of Convolutional Neural Networks (CNNs) to enable automated detection of geometric features from CAD models and scanned parts and to extract GD&T symbols from 2D and 3D drawings to assess tolerance zones [30]. Another important contribution of IA in tolerance analysis is related to the use of reinforcement learning and genetic algorithms to optimize tolerance allocation by reducing manufacturing costs while maintaining quality [31]. In the context of Additive Manufacturing, AI-driven optimization methods could help compensate for material shrinkage, warping, and roughness, ensuring more reliable dimensional control. Finally, deep learning models integrated with Control Measurement Machines and optical inspection systems can analyze dimensional deviations in real time, improving inspection speed and reliability [32]. AI-driven anomaly detection techniques can also assist in predicting potential defects before they occur. Despite its advantages, demonstrated by scientific literature, AI-driven GD&T faces challenges such as data availability, model interpretability, and integration with existing manufacturing workflows. Accordingly, future research should focus on developing explainable AI models, improving AI robustness, and standardizing AI-based GD&T methodologies.

6. Standardization of Inspection for AM

As briefly aforementioned, complex AM geometries may not always be appropriate for conventional inspection instruments (CMMs, calipers). As a result, non-contact measuring techniques for AM component inspection will be more frequently included in GD&T standards [33].

These methods include the following: (1) using laser or structured light scanners to provide accurate digital representations of the surface geometry of an item, allowing for thorough surface examinations; (2) optical 3D scanning, which projects fringe patterns onto a component's surface and uses cameras to record the reflected data to perform precise shape and dimension analyses without physical contact; (3) Computed Tomography (CT) scanning, which uses X-rays to capture internal and external geometries and enables both the identification of subsurface defects and the dimensional assessment of internal features that are otherwise unmeasurable with traditional methods. Therefore, standardization and benchmarking in GD&T for AM will increasingly be involved together with the mature development of Reverse Engineering techniques, allowing for the development of uniform methods for defining, measuring, and verifying tolerances specific to AM processes. These efforts will ensure consistency, interoperability, and reliability across industries using AM technologies. In general, there is a lack of standardization in terms of tolerancing dimensional stability for AM parts. This is mainly due to the rising of AM process-dependent deviations on manufactured parts, due to thermal distortions, warping, and shrinkage, depending on the different AM technologies and on the wide range of materials used to produce the parts. To confront these issues, some efforts are currently in development, based on the standard ISO/ASTM 52910:2018 [34], providing a set of guidelines and recommendations for using AM in product design. However, more studies are expected in the future regarding this important topic.

7. GD&T Adaptation to Different Technologies

GD&T needs also to be adapted to different processes commonly adopted to manufacture mechanical parts, such as Powder Bed Fusion (PBF) and Directed Energy Deposition (DED), while also considering the unique behaviors of materials like metals, polymers, and ceramics. In PBF processes, dimensional accuracy is influenced by powder particle size, laser scanning strategies, and thermal distortions, requiring tighter controls on form tolerances (flatness, cylindricity) and position tolerances (true position, concentricity) [35]. DED processes, on the other hand, produce larger, near-net-shape components with higher thermal input, often leading to higher residual stress and lower feature resolution, making profile tolerances more critical for accounting for surface variations [36]. Because of anisotropic shrinkage and microstructural alterations, metal AM parts require process-specific tolerance compensation. With an emphasis on profile and straightness tolerances, polymer AM, like FDM or SLS, necessitates GD&T modifications for warping and layer bonding irregularities. Because of its brittleness and sintering shrinkage, ceramic AM, which finds utility in biomedical and aerospace applications, poses difficulties that call for GD&T solutions that take post-processing dimensional changes into account. By tailoring GD&T to each AM process and material, functional accuracy is guaranteed, post-processing requirements are decreased, and reliability is improved across a range of applications.

8. GD&T in AM Assembly Parts

As widely recognized, GD&T plays a crucial role in ensuring the dimensional and geometric accuracy of assemblies by allowing the control of both dimensional and geometric tolerances by leveraging ISO 2692:2001 [37] and ASME Y14.5 - 2018 (R2024) [38] rules listed in Table 1.

Table 1. ISO [37] and ASME [38] rules for assemblies.

ISO Rules	ASME Rules
Rule 1: Independence Principle Each dimensional or geometric prescription must be respected in itself (independently) unless a particular relationship is specified	Rule 1: Envelope requirement When specifying only dimensional tolerances, these always apply under maximum envelope conditions [equivalent to ISO Rule 2]. No modifier is used.
Rule 2: Envelope requirement The shape of an element must never exceed the envelope of the perfect shape corresponding to the maximum material condition. ISO adopts modifier E	Rule 2: Interdependence To establish an interdependence between geometric and dimensional tolerances, a modifier has to be adopted according to the principle of Maximum Material or Minimum Material [equivalent to ISO Rule 3]. ASME adopts modifiers M and L
Rule 3: Interdependence To establish an interdependence between geometric and dimensional tolerances, a modifier has to be adopted according to the principle of Maximum Material or Minimum Material. ISO adopts modifiers M and L	Rule 3: Regardless of Feature Size Each dimensional or geometric prescription must be respected in itself (independently) unless a particular relationship is specified. ASME adopts modifier RFS

The aforementioned rules are clearly applicable to 2D drawings, ensuring the compliance of the manufactured parts to be assembled with their designed functional scope. Since AM parts are purely model-driven manufacturing processes [39], the role of 2D drawings in designing AM parts is of little value. Therefore, GD&T annotation practices should move toward 3D annotations. This is a big challenge in the GD&T field, especially due to complex geometries, e.g., for Topologically Optimized parts. For these parts, geometric and dimensional annotations may not be able to capture either the complexity of the manufactured surfaces or the surface of mating features. To partially solve this issue, GD&T practices are moving from purely symbolic definition to semantic CAx-integrated annotation, the latter being this intended for computer-based use.

9. Discussion and Conclusions

The integration between Geometric Dimensioning and Tolerancing and Additive Manufacturing is a critical aspect of ensuring the reliability, consistency, and functionality of AM-produced components. Unlike subtractive manufacturing, where the manufacturing process is relatively simpler, AM throws many challenges, including complex freeform geometries, anisotropic material properties, and process-induced deviations that are not easily implemented in the GD&T standard. All these issues indicate a shift from traditional GD&T approaches toward more flexible and encompassing methodologies.

The first important development is the adaptation of GD&T principles to AM-specific geometries, where functional datum systems and freeform surface tolerancing are gaining prominence. Because layer-by-layer fabrication creates distinct roughness patterns and dimensional irregularities, it is also important to incorporate tolerance stack-up models and surface texture analysis specifically designed for AM. Additionally, the integration of Model-Based Definition (MBD) into the digital thread seems to be a challenging yet effective strategy for ensuring continuity from design to manufacturing by reducing interpretation mistakes and enabling automated tolerance checking. Again, the key to guaranteeing AM's industrial scalability includes standardization and benchmarking. Not by chance, standardized procedures for the inspection, tolerancing, and quality control of AM components are being developed by organizations like ASME and ISO. Finally, a significant step forward is the increasing reliance on non-contact inspection techniques, which allows for the verification of complex internal and external geometries.

However, challenges still exist. The development of interoperability among various AM procedures, software tools, and GD&T standards continues to evolve and calls for constant improvement, as well as industry cooperation. Furthermore, current GD&T frameworks do not adequately handle the additional tolerancing issues presented by multi-material and functionally graded AM components. It is also important to note that when a given AM process cannot ensure a set of tolerances that comply with the functional requirements of a part, AM either cannot be selected as an appropriate process, or it needs to be followed by another additional manufacturing process. This is particularly true for complex assemblies consisting of many mated parts, where strict control of the tolerance chain is required in addition to the tolerances of the specific parts.

Future work must concentrate on integrating GD&T with AM-driven design paradigms as research advances, making sure that tolerancing strategies complement AM's special strengths rather than imposing limitations inherited from conventional manufacturing techniques. Overall, GD&T and AM integration is a developing topic that necessitates interdisciplinary collaboration between standards organizations, software developers, and engineers. GD&T can fully support the expanding industrial usage of AM by developing digital and, why not, AI-driven quality control procedures, improving tolerance definitions for AM-specific geometries, and setting internationally recognized benchmarks. The ultimate objective is to provide a strong, automated, and scalable framework that guarantees AM-produced parts meet functional and regulatory criteria while preserving manufacturing efficiency and cost-effectiveness. This, to the best of the author's knowledge, is the future of engineering practice, where traditional manufacturing methods will be increasingly integrated with additive techniques and necessarily connected with the problems inherent to the need to tolerate mechanical components with an unambiguous and consistent system. For this reason, this Perspective Paper aims to inspire all researchers and practitioners working in the mechanical engineering sector to implement every possible effort to promote the integration of GD&T and AM as a strategy for the effective manufacturing of mechanical components and complex mechanisms.

Funding: This research received no external funding.

Data Availability Statement: No data are available for this research.

Conflicts of Interest: The author declares no conflicts of interest.

References

1. Aghaei, S.; Mackieson, B. Digital GD&T Control Features Validation from 3D CAD to Metrology Software. *Meas. Sens.* **2024**, 101495. [CrossRef]
2. Pathak, V.K.; Singh, A.K.; Sivadasan, M.; Singh, N.K. Framework for Automated GD&T Inspection Using 3D Scanner. *J. Inst. Eng. (India) Ser. C* **2018**, *99*, 197–205.
3. Furferi, R. Tolerance Analysis Methods for the Application of ISO and ASME GD&T to Mechanical Component: 2D and 3D Case Studies. *J. Adv. Manuf. Sci. Technol.* 2025; 2025020, *ahead-of-print*. [CrossRef]
4. Yang, S.; Zhao, Y.F. Additive Manufacturing-Enabled Design Theory and Methodology: A Critical Review. *Int. J. Adv. Manuf. Technol.* **2015**, *80*, 327–342. [CrossRef]
5. Pei, E.; Kabir, I.; Breški, T.; Godec, D.; Nordin, A. A Review of Geometric Dimensioning and Tolerancing (GD&T) of Additive Manufacturing and Powder Bed Fusion Lattices. *Prog. Addit. Manuf.* **2022**, *7*, 1297–1305.
6. Rupal, B.S.; Qureshi, A.J. Geometric Deviation Modeling and Tolerancing in Additive Manufacturing: A GD&T Perspective. In Proceedings of the 1st Conference of NSERC Network for Holistic Innovation in Additive Manufacturing (HI-AM), Waterloo, ON, Canada, 22–23 May 2018; pp. 1–6.
7. Kabir, I. Fundamentals Matter: Why GD&T Remains Essential in Modern Manufacturing. *Quality* **2024**, *63*, 9.
8. Xiao, J.; Anwer, N.; Durupt, A.; Le Duigou, J.; Eynard, B. Information Exchange Standards for Design, Tolerancing and Additive Manufacturing: A Research Review. *Int. J. Interact. Des. Manuf. (IJIDeM)* **2018**, *12*, 495–504. [CrossRef]

9. Maupou, C.; Yang, Y.; Fodop, G.; Qie, Y.; Migliorini, C.; Mehdi-Souzani, C.; Anwer, N. Automatic Raster Engineering Drawing Digitisation for Legacy Parts Towards Advanced Manufacturing. *Procedia CIRP* **2024**, *129*, 234–239. [CrossRef]
10. Abdolahi, A.; Soroush, H.; Khodaygan, S. Process Capability Analysis of Additive Manufacturing Process: A Machine Learning-Based Predictive Model. *Rapid Prototyp. J.* 2025; *ahead-of-print*. [CrossRef]
11. Zhang, Q.; Liu, J.; Chen, X. A Literature Review of the Digital Thread: Definition, Key Technologies, and Applications. *Systems* **2024**, *12*, 70. [CrossRef]
12. Gregory, M.; Vinsevich, D. Defining and Tolerancing Freeform Surfaces for Manufacturing. *Proc. SPIE* **2024**, *13133*, 131330A. [CrossRef]
13. Fan, Y.; Nikolov, D.K.; Pomerantz, M.; Vamivakas, A.N.; Rolland, J.P. Freeform Manufacturing Process Optimized by Automatically Generated CAD Model. *Proc. SPIE* **2024**, *12798*, 127982C. [CrossRef]
14. *ISO 21920-1:2021*; Geometrical product specifications (GPS)—Surface texture: Profile—Part 1: Indication of surface texture. International Organization for Standardization: Geneva, Switzerland, 2021.
15. *ASME Y14.46*; Product Definition for Additive Manufacturing. American Society of Mechanical Engineers: New York, NY, USA, 2022.
16. Calignano, F.; Mercurio, V. Influence of the Aspect Ratio of an Additive Manufacturing Component on the Surface Roughness during Cutting with Wire EDM. *Int. J. Interact. Des. Manuf. (IJIDeM)* **2024**, *18*, 1075–1085. [CrossRef]
17. Jiang, J.; Xu, X.; Stringer, J. Support Structures for Additive Manufacturing: A Review. *J. Manuf. Mater. Process.* **2018**, *2*, 64. [CrossRef]
18. *ISO/ASTM 52920:2023*; Additive Manufacturing—Qualification Principles—Requirements for Industrial Additive Manufacturing Processes and Production Sites. International Organization for Standardization: Geneva, Switzerland, 2023.
19. Vora, H.D.; Sanyal, S. A Comprehensive Review: Metrology in Additive Manufacturing and 3D Printing Technology. *Prog. Addit. Manuf.* **2020**, *5*, 319–353.
20. Buj-Corral, I.; Sánchez-Casas, X.; Luis-Pérez, C.J. Analysis of AM Parameters on Surface Roughness Obtained in PLA Parts Printed with FFF Technology. *Prog. Addit. Manuf.* **2021**, *13*, 2384.
21. Gockel, J.; Sheridan, L.; Koerper, B.; Whip, B. The Influence of Additive Manufacturing Processing Parameters on Surface Roughness and Fatigue Life. *Int. J. Fatigue* **2019**, *124*, 380–388.
22. Uccheddu, F.; Carfagni, M.; Governi, L.; Furferi, R.; Volpe, Y.; Nocerino, E. 3D Printing of Cardiac Structures from Medical Images: An Overview of Methods and Interactive Tools. *Int. J. Interact. Des. Manuf. (IJIDeM)* **2018**, *12*, 597–609.
23. Buonamici, F.; Furferi, R.; Governi, L.; Lazzeri, S.; McGreevy, K.S.; Servi, M.; Talanti, E.; Uccheddu, F.; Volpe, Y. A CAD-Based Procedure for Designing 3D Printable Arm-Wrist-Hand Cast. *Comput. Aided Des. Appl.* **2019**, *16*, 25–34.
24. Herderick, E.D. Progress in Additive Manufacturing. *JOM* **2015**, *67*, 580–581.
25. Ameta, G.; Witherell, P. Representation of Graded Materials and Structures to Support Tolerance Specification for Additive Manufacturing Application. *J. Comput. Inf. Sci. Eng.* **2019**, *19*, 021008. [CrossRef]
26. *ISO 16792:2021*; Technical product documentation—Digital product definition data practices. International Organization for Standardization: Geneva, Switzerland, 2021.
27. *EN ISO 1101:2017*; Geometrical Product Specifications (GPS)—Geometrical Tolerancing—Tolerances of Form, Orientation, Location and Run-Out. International Organization for Standardization: Geneva, Switzerland, 2017.
28. Liu, C.; Tian, W.; Kan, C. When AI Meets Additive Manufacturing: Challenges and Emerging Opportunities for Human-Centered Products Development. *J. Manuf. Syst.* **2022**, *64*, 648–656.
29. Zhu, Z.; Anwer, N.; Huang, Q.; Mathieu, L. Machine Learning in Tolerancing for Additive Manufacturing. *CIRP Ann.* **2018**, *67*, 157–160. [CrossRef]
30. Lin, Y.H.; Ting, Y.H.; Huang, Y.C.; Cheng, K.L.; Jong, W.R. Integration of Deep Learning for Automatic Recognition of 2D Engineering Drawings. *Machines* **2023**, *11*, 802. [CrossRef]
31. Ghali, M.; Elghali, S.; Aifaoui, N. Genetic Algorithm Optimization Based on Manufacturing Prediction for an Efficient Tolerance Allocation Approach. *J. Intell. Manuf.* **2024**, *35*, 1649–1670. [CrossRef]
32. Cheng, Z.Y.; Sun, Y.; Hu, K.; Li, J.; Lu, T.F.; Li, R.J. Deep Learning-Based Intelligent Measurement Methods and System for CMM. *Measurement* **2023**, *221*, 113474. [CrossRef]
33. Boeckmans, B.; Probst, G.; Zhang, M.; Dewulf, W.; Kruth, J.P. ISO 10360 Verification Tests Applied to CMMs Equipped with a Laser Line Scanner. In *Dimensional Optical Metrology and Inspection for Practical Applications V*; SPIE: Bellingham, WA, USA, 2016; Volume 9868, pp. 30–39.
34. *ISO/ASTM 52910:2018*; Additive Manufacturing—Design—Requirements, Guidelines and Recommendations. International Organization for Standardization: Geneva, Switzerland, 2018.
35. Xiao, X.; Roh, B.M.; Hamilton, C. Porosity management and control in powder bed fusion process through process-quality interactions. *CIRP J. Manuf. Sci. Technol.* **2022**, *38*, 120–128. [CrossRef]

36. Dominguez, L.A.; Xu, F.; Shokrani, A.; Flynn, J.M.; Dhokia, V.; Newman, S.T. Guidelines when considering pre & post processing of large metal additive manufactured parts. *Procedia Manuf.* **2020**, *51*, 684–691.
37. *ISO 2692:2021*; Geometrical Product Specifications (GPS)—Geometrical Tolerancing—Maximum Material Requirement (MMR), Least Material Requirement (LMR) and Reciprocity Requirement (RPR). International Organization for Standardization: Geneva, Switzerland, 2021.
38. *ASME Y14.5-2018 (R2024)*; Dimensioning and Tolerancing. American Society of Mechanical Engineers: New York, NY, USA, 2018.
39. Witherell, P.; Herron, J.; Ameta, G. Towards Annotations and Product Definitions for Additive Manufacturing. *Procedia CIRP* **2016**, *43*, 339–344. [CrossRef]

applied sciences

MDPI

Article

Comparing Bolt Implementation Methods in Vibration Testing for Accurate Dynamic Behavior Analysis

Jong Hyuk Back [1], Jeong Bin Bae [1], Ji Hye Kang [2] and Jung Jin Kim [1,*]

[1] Department of Mechanical Engineering, Keimyung University, Daegu 42601, Republic of Korea; bigbell608@gmail.com (J.H.B.); willy600301@gmail.com (J.B.B.)
[2] SL Corporation, Gyeongsan-si 38470, Republic of Korea; kangjihye@slworld.com
* Correspondence: kjj4537@gmail.com; Tel.: +82-53-580-5290; Fax: +82-53-715-2029

Abstract: Vibration testing is crucial for understanding structural dynamics, yet conventional modeling of bolt connections often leads to significant inaccuracies. This study systematically compares six bolt connection methods—bonded, adaptive bonded, joint, beam, screw, and fixed bolt—using a finite element analysis of a headlamp vibration test jig. The six bolt connection methods were selected based on approaches adopted in previous studies. The experimental results identified the joint connection method as the most accurate, minimizing deviations in natural frequency to 7.6 Hz compared to experimental tests at 493.2 Hz, while bonded methods overestimated the frequency at 544.1 Hz due to excessive stiffness assumptions. Efficiency analyses highlighted bonded methods as the most computationally streamlined, offering preprocessing times as short as 30 s and shorter overall analysis times. These findings emphasize the importance of selecting appropriate bolt connection methods in the early design phase to ensure accurate natural frequency predictions and mode shape representations. Although this study does not consider bolt preload forces, the work shows the possibility of offering practical guidelines for improving the reliability and efficiency of vibration test jig designs by bridging the gap between analysis and experimental results.

Keywords: bolt implementation; dynamic behavior analysis; vibration testing; finite element analysis

1. Introduction

Vibration testing provides critical data for assessing the safety and durability of a structural design by analyzing the natural frequencies and mode shapes of the structure [1,2]. This analysis is crucial for understanding the dynamic behavior of structures and assessing design suitability and durability [3–5]. Resonance at specific frequencies, in particular, can increase component fatigue and reduce durability, making design review and vibration testing an essential preventative step [6–8]. These vibration tests are performed by fixing the structure under test. A jig is required to obtain accurate test results with natural frequencies and mode shapes superior to those of the test object [9]. This is because the influence of the jig's properties on the test results is minimized, enabling an accurate evaluation of the actual durability performance of the structure [10].

Vibration test jigs are typically assembled using ribs and plates fastened with bolts. The bolt connection method significantly affects properties such as rigidity and natural frequency [11–13]. In the early design stages, bolt connections are often simplified and modeled as bonded (perfect bonding) to improve analysis efficiency. Although this approach reduces analysis time, it fails to fully capture the actual contact behavior of bolt connections, leading to discrepancies between the analysis and test results [14–16]. This hinders the derivation of accurate test results and increases the likelihood that the jig design properties will not meet expectations under real-world conditions.

The bolt connection method significantly affects the natural frequencies and mode shapes of a vibration test jig. Therefore, differences may arise between the analysis and experimental results depending on the contact conditions and connection method chosen [11]. Numerous studies have analyzed the effects of bolt connection methods on the natural frequencies and mode shapes of structures through experiments and numerical analysis, emphasizing their critical role in ensuring design accuracy and appropriateness [17,18]. For instance, a study showed that failing to reflect the actual contact conditions of the bolts during vibration testing can reduce the consistency between the analysis and experimental results, highlighting the importance of proper bolt implementation methods to improve design accuracy [19]. Additionally, a study that experimentally demonstrated the impact of bolt connection methods on natural frequencies and mode shapes suggested that selecting the correct connection method during the initial design stage is essential [20].

To address the discrepancy between analysis and test results, various bolt implementation methods have been proposed [21]. For example, connecting a line body under rigid conditions can improve analysis efficiency by providing shorter analysis times in vehicle engine vibration testing [22]. Furthermore, the simplified connection approach has been effectively applied in the vibration analysis of diesel engine cylinders, streamlining bolt implementation while maintaining adequate accuracy [23]. However, research on bolt implementation methods in vibration testing remains fragmented, and a systematic comparison and evaluation of the performance and limitations of each method are lacking.

This study analyzes the natural frequencies, mode shapes, and efficiency of various bolt implementation methods using a vibration test jig to determine the most suitable method for design and analysis. To achieve this, a headlamp jig and M6 standard tapped bolts currently in use were selected as the research materials. The bolts were modeled using six different implementation methods, and their natural frequencies, mode shapes, analysis times, and preprocessing efficiency were compared. Additionally, the natural frequencies obtained from vibration tests were compared with those of the analysis models to evaluate the effectiveness of each method. Through this analysis, this study seeks to provide practical guidelines for design and analysis stages, contributing to improved accuracy and efficiency in structural design.

2. Materials and Methods

2.1. Specifications

In this study, M6 standard tapped bolts were selected as the research model. These bolts are used in actual vibration tests to connect the ribs and feature a design that forms internal threads directly on the counterpart without requiring nuts, particularly in situations where through-holes cannot be drilled. Figure 1 and Table 1 present the dimensions and specifications of the M6 bolts used in this study.

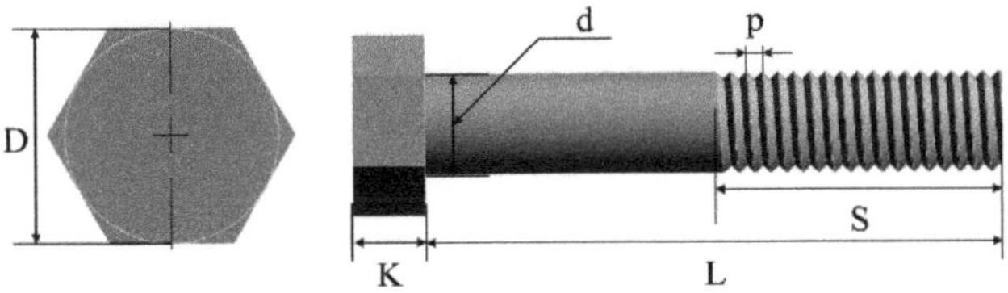

Figure 1. M6 bolt dimensions.

Table 1. M6 bolt dimensions.

	Length [L]	Diameter [d]	Head Diameter [D]	Head Thickness [K]	Pitch [p]	Threaded Length [S]
Dimension (mm)	40	6	10	4	1	20

Figure 2 shows the headlamp vibration test jig used for analysis and testing. Headlamps are continuously exposed to vibration and shocks during operation, making vibration testing essential. Furthermore, the headlamp vibration test jig, which requires numerous bolt fastenings, is well-suited for comparing behaviors based on bolt fastening methods. The jig weighed 67.8 kg and comprised 14 ribs, 2 lamp-fastening sections, and 1 lower plate. The ribs were 20 mm thick, while the lower plate was 25 mm thick, with dimensions of 892.4 mm along the x-axis and 652.7 mm along the y-axis. The jig's overall height, measured from the top of the lower plate, was 628.8 mm. A total of 291 bolt holes were created on the jig at 30 mm intervals, with additional holes at 100 mm intervals on the lower plate for attachment to the vibration testing machine (Figure 2).

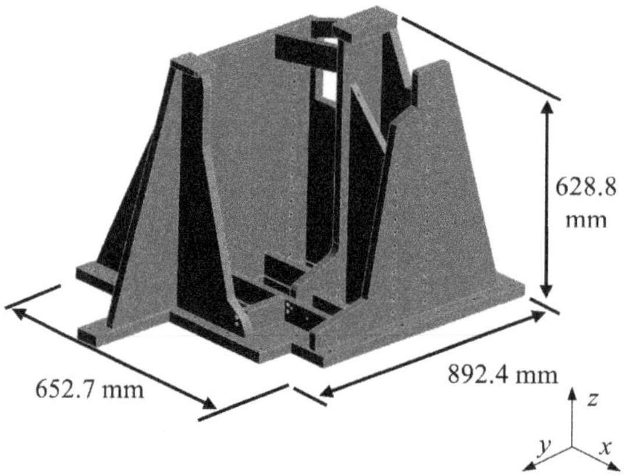

Figure 2. Structure and fastening positions of the vibration test jig for a headlamp.

2.2. Finite Element Analysis

In this study, six methods were used to simulate bolt connections: bonded, adaptive bonded, joint, beam, screw, and fixed bolt. The six selected methods in this study represent the functions for a bolt simulation available in ANSYS and Inspire (i.e., widely used commercial software). The adaptive bonded, joint, and beam methods were provided in ANSYS, while the screw and fixed bolt methods were provided in Inspire. Particularly, the adaptive bonded and beam methods were used in previous studies, such as those that analyzed diesel engine vibrations [23], and the joint method was also frequently applied in engine vibration testing research [22]. On the other hand, the screw and fixed bolt methods were included in this study for comparative purposes although they were not widely studied. Figure 3 shows the fastening appearance of the bolts used. In addition, Figure 4 shows the application sequence of the proposed method in a flow chart.

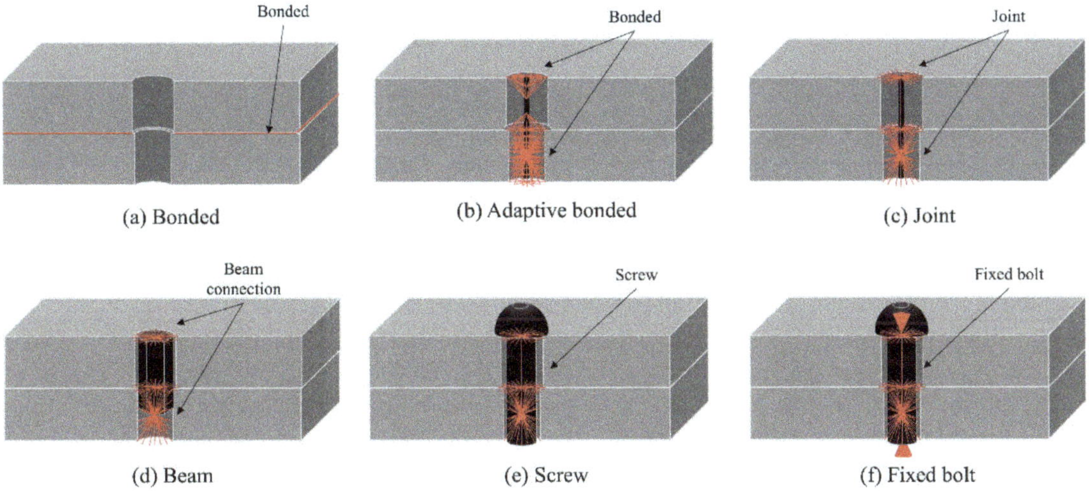

Figure 3. Implementation methods for bolt connections.

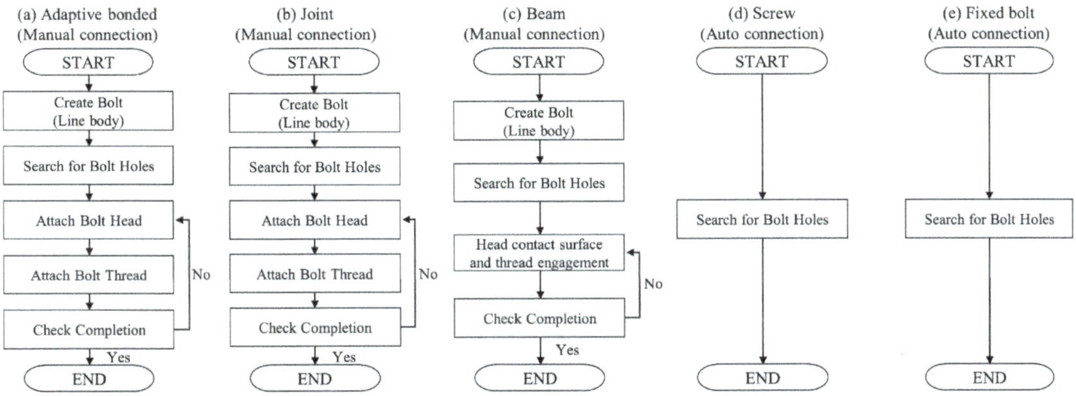

Figure 4. Implementation procedures for bolt connections.

The bonded connection method (Figure 3a) establishes a fully bonded state by applying bonded conditions to the contact area without modeling the bolt. Contact conditions are applied to prevent deformation or slippage on all contact surfaces, ensuring a solid bond between structures. This method assumes that the entire joint area acts as a rigid body, regardless of the contact area, thus representing high bonding strength. This connection method was implemented using the "Contacts" tool in ANSYS Connections.

The adaptive bonded connection method (Figure 3b) models bolts as line bodies, assigning contact areas and bonded conditions to connect them. To implement this, bolts are created by connecting two line bodies. These line bodies are divided into threaded and non-threaded sections for representation. At the non-threaded sections, contact conditions are applied to the upper and lower edges of the holes to simulate a weaker bonding strength. Additionally, contact conditions are applied between the line bodies and surfaces with internal threads to represent the threaded contact. This connection method was implemented using the "Contacts" tool in ANSYS Connections.

The joint connection method (Figure 3c) models bolts as line bodies and rigidly connects them with the contact areas. Bolts were created as line bodies in the same way as in the adaptive bonded method. For contact representation, the non-threaded sections of the line body were assigned contact conditions at the upper and lower edges. Additionally, the threaded sections were assigned contact conditions between the line body and the internal thread surfaces. This method was implemented using the "Joint" tool in ANSYS Connections.

The beam connection method (Figure 3d) implements bolts as virtual beam elements and rigidly connects the contact areas. A notable feature of this method is that it does not require prior creation of bolt bodies. Unlike the adaptive bonded method, where bolt contacts are specified separately, the creation of beams directly serves as the bolt representation. Therefore, contact conditions are directly applied to the edge of the surface where the bolt head is fastened and the area on the lower plate where the internal threads are located. This method was implemented using the "Beam" tool in ANSYS Connections.

The screw connection method (Figure 3e) models bolts as line bodies and connects the contact areas using Rigid Body Element 2 (RBE2) constraints. RBE2 ensures that all connected nodes behave as a rigid body, maintaining identical displacements and rotations across the nodes. In the screw connection method, bolts could be created simply without requiring additional contact conditions or the need to locate bolt positions. This method simplifies fastening, enabling efficient modeling and analysis. This connection method was implemented using the "Fastener" tool in the Inspire analysis program.

The fixed bolt connection method (Figure 3f) also models bolts as line bodies and connects the contact areas using RBE2 constraints. Unlike the screw connection method, the nodes at the top and bottom of the CBAR are designated as fixed points, ensuring that both ends of the bolt remain completely fixed. Similar to the screw connection method, the fixed bolt method allows bolts to be created simply without assigning additional contact conditions or locating bolt positions. The fixed bolt method provides a fastening method that simulates connections where rigidity is relatively important. This connection method was implemented using the "Fastener" tool in the Inspire analysis program.

For the representation of line bodies, BEAM188 elements were used in ANSYS analysis, while CBAR elements were used in Inspire. BEAM188 elements have the characteristic of allowing thickness information to be input for one-dimensional structures. Additionally, these elements have six degrees of freedom and can account for axial forces, shear forces, bending, and torsion moments. CBAR elements have the same degrees of freedom and can similarly represent the physical properties of bolts.

In general, computational analysis results tend to indicate higher natural frequencies compared to experimental values, as they cannot perfectly account for all conditions. To address this, contact friction between contact areas, excluding bolt joints, was not considered in the contact conditions. This assumption was based on the relatively small contact area between the ribs. In addition, the bolts, positioned at 30 mm intervals, were considered sufficient to provide the necessary structural strength for the connection [24]. Moreover, the preload of the bolts was excluded, as it could excessively influence the computational analysis results by further increasing the structural strength.

Figure 5 shows the finite element model of the entire jig. The FE model of the jig was constructed using SOLID185 in ANSYS, 8-node solid elements. This element uses the simplified enhanced strain formulation. Enhanced strain formulation modifies gradient tensor and strains in lower-order elements to prevent shear and volumetric locking [25]. The model consisted of 1,094,136 nodes and 627,725 elements, with an average aspect ratio of 2.31 and an average Jacobian ratio of 0.97, indicating sufficient mesh quality [26]. The detailed calculations are provided in Sections A and B. The bolt material properties were assigned based on AISI 304 stainless steel (density: 8×10^{-6} kg/mm^3, Young's modulus: 195.0 GPa, Poisson's ratio: 0.29, and yield strength: 215 MPa). The jig material properties were assigned based on Aluminum Die Casting (ALDC) (density: 2.75×10^{-6} kg/mm^3,

Young's modulus: 69.6 GPa, Poisson's ratio: 0.33, and yield strength: 165 MPa). The analysis was conducted using the ANSYS 24 R1 and Inspire 2023 programs. A personal computer (Intel i9-12900K, GPU: GEFORCE RTX 3080, RAM: 64 GB, Santa Clara, CA, USA) was used for the analyses.

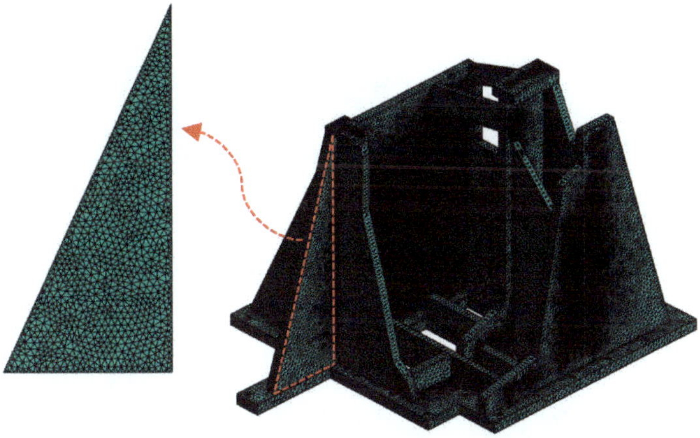

Figure 5. Finite element model of the jig.

2.3. Testing Methodology

The vibration test was conducted using a Dongling's ES-200LS2-650 vibration tester (Suzhou, China). This equipment provides a maximum impact force of 600 kN, a frequency range of 5–2100 Hz, a maximum acceleration of 1000 m/s^2 (100 g), a maximum displacement of ±76 mm, and a maximum allowable weight of 2500 kg. Acceleration measurements were recorded using Kistler's 8703A250M5 accelerometer (Winterthur, Switzerland). Data acquisition and analysis were performed using the VR9500 Vibration Controller along with the VibrationVIEW software (v.15.2, Vibration Research, Jenison, MI, USA). The test was conducted over a frequency range of 30 Hz to 700 Hz, which is set to be more than twice the minimum criterion for the first natural frequency of the vibration test jig (350 Hz). In addition, the vibration test was conducted along a single axis in the X-axis direction [27,28].

Figure 6 shows the jig fixing position and the position where the acceleration sensor is attached in the vibration experiment. The test subject was the headlamp vibration test jig, with a total weight of 70.4 kg for the jig and bolts combined. Sensors were fixed using adhesive to the lamp-fastening position (Top Head) at the topmost part of the vibration test jig (Figure 6). The sensors were positioned at the topmost part of the vibration test jig (Top Head) to sensitively reflect the overall vibration characteristics of the structure. This is because the top section experiences the maximum deformation and acceleration during resonance, making it the optimal location for analyzing natural frequencies and mode shapes [29]. The jig was made of ALDC material, and the test subject was fastened using M6 standard tapped bolts. The jig was secured to the vibration tester using M10 standard hex wrench-tapped bolts, with the fixed location set at the center of the top plate of the vibration tester. The test results derived the first natural frequency by analyzing changes in acceleration and phase angle. Resonance frequency was considered to occur when the change in phase angle exceeded 90°.

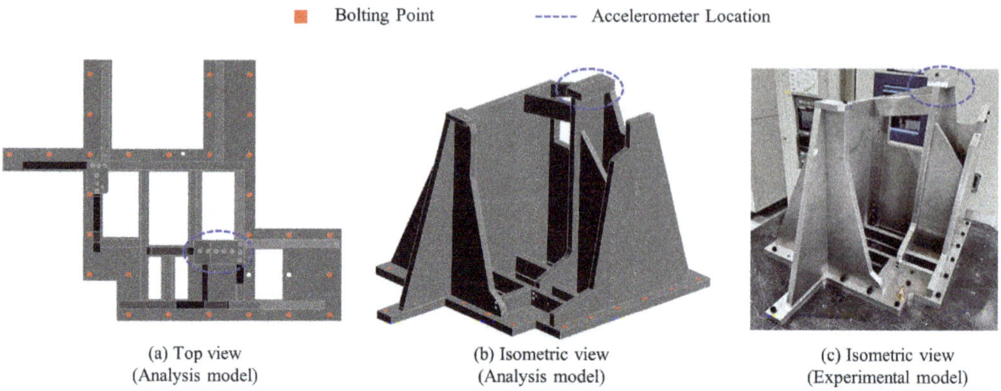

Figure 6. Vibration testing equipment and jig: (**a**) top view, (**b**) isometric view (analysis model), (**c**) isometric view (experimental model).

2.4. Comparison of Analysis and Test Results

This study was conducted to identify the optimal bolt connection method by comparing the natural frequencies and mode shapes observed in the jig for various bolt connection methods against vibration test results. The comparison was conducted by evaluating the representativeness of the bonded connection method in vibration testing. It involved analyzing changes in the jig's natural frequencies and mode shapes depending on the connection method, assessing the similarity with test results, and performing an efficiency analysis.

First, the study evaluated whether the bonded connection method, which is the most common method used to represent bolt connections, effectively represents vibration testing. For this purpose, the jig's natural frequencies derived from vibration testing were compared with those obtained through bonded-based vibration analysis. Here, natural frequency refers to the frequency at which resonance occurs at the Top Head of the jig (493.2 Hz).

Second, the study analyzed whether there were similarities or differences in the jig's natural frequencies and mode shapes depending on the bolt connection method. For this analysis, the first natural frequencies and mode shapes of the jig were compared for each bolt connection method [12,30,31]. The first natural frequency is the most sensitive indicator of structural stiffness and directly reflects changes in bonding strength and stiffness caused by different connection methods. Additionally, the first natural frequency represents the overall dynamic characteristics of the structure and serves as a critical criterion in design and analysis [32–34].

Third, the bolt connection method that produces natural frequencies most similar to those of the vibration test was determined. For this, the natural frequencies of the Top Head measured during vibration testing were compared with those calculated for each bolt connection method. Through this comparison, the connection method showing natural frequencies closest to the test results was selected.

Fourth, the efficiency of the analysis for each bolt connection method was compared. For this purpose, the CP Time and total memory usage required for the analysis of each connection method were analyzed. CP Time represents the total processing time the CPU spends on actual computation during the analysis and reflects the complexity of an analysis model and the efficiency of computational tasks. The total memory usage serves as a metric to assess the complexity and resource consumption of the analysis model. High memory requirements can slow down the analysis or even cause it to fail, making memory usage an important metric. Additionally, the time and effort required for generating finite element models of bolts and the fastening procedures during preprocessing were also analyzed to comprehensively evaluate the usability of each connection method.

3. Results

Figure 7 shows the changes in acceleration and phase angle measured in the jig during the vibration test. The first natural frequency of the jig is 493.2 Hz, with an acceleration peak of 17.1 g and a phase angle change of 85° at this frequency. This result varied significantly from the analysis result of the bonded connection method. The natural frequency of the jig with the bonded connection method was 544.1 Hz, approximately 10.3% higher than the vibration test result.

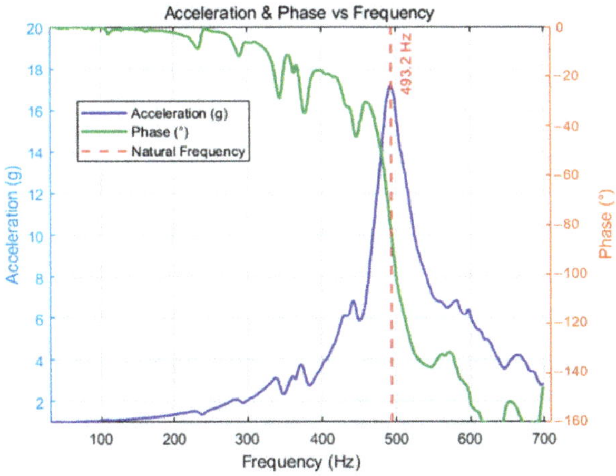

Figure 7. Acceleration and phase response with natural frequency identification.

Figure 8 shows the natural frequencies and mode shapes observed in the first mode for different bolt connection methods. The colors in Figures 8 and 9 indicate the relative mode shape deformations across the structure. Four methods (adaptive bonded, beam, screw, joint) exhibited lower natural frequencies and distinct mode shapes compared to the bonded connection method. For example, the adaptive bonded method showed a local bending mode (Local Mode) along the y-axis in the ribs, with a natural frequency of 324.7 Hz. The beam method also exhibited local bending along the y-axis in the ribs, with the lowest natural frequency of 301.8 Hz. The screw method showed local bending in the ribs, with a natural frequency of 388.2 Hz. The joint method exhibited bending deformation along the x-axis at the Top Head, with a natural frequency of 500.8 Hz. By contrast, the fixed bolt method showed a higher natural frequency (881.9 Hz) compared to the bonded connection method, with a local mode displaying bending along the y-axis in specific ribs.

Figure 9 illustrates the mode shapes and natural frequencies observed at the Top Head for different bolt connection methods. The Top Head position was specifically selected because it corresponds to the location where the accelerometer sensor was attached during the vibration test. This verifies that the analysis aligns with the actual experiment setup. The bolt connection method with the highest frequency was the bonded connection method, which exhibited a frequency of 544.1 Hz in the first mode. The method with the lowest frequency was the beam connection method, which showed a frequency of 349.8 Hz in the fifth mode. However, the bolt connection method with results most similar to those of the vibration test was the joint connection method, which exhibited a frequency of 500.8 Hz in the first mode, showing only a 7.6 Hz difference from the test result. The fixed bolt connection method, which exhibited the highest frequency in the first mode, did not show a mode occurring at the Top Head among the six modes analyzed.

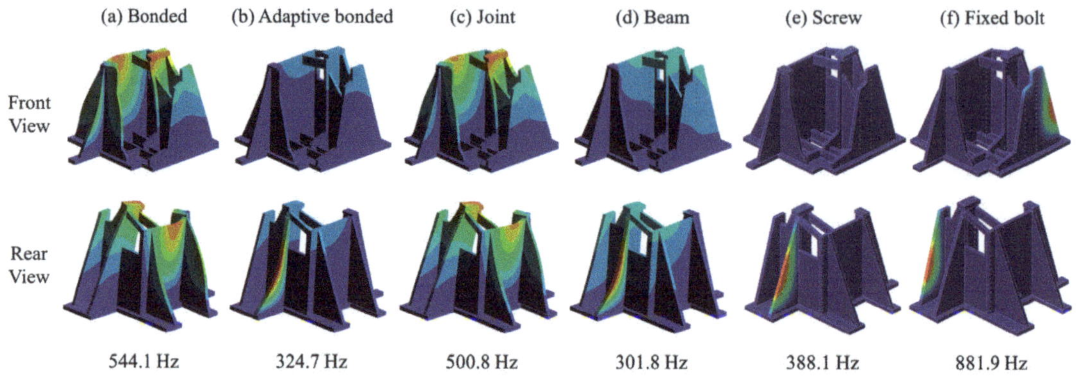

Figure 8. Natural frequencies and mode shapes based on different bolt connection methods: First mode.

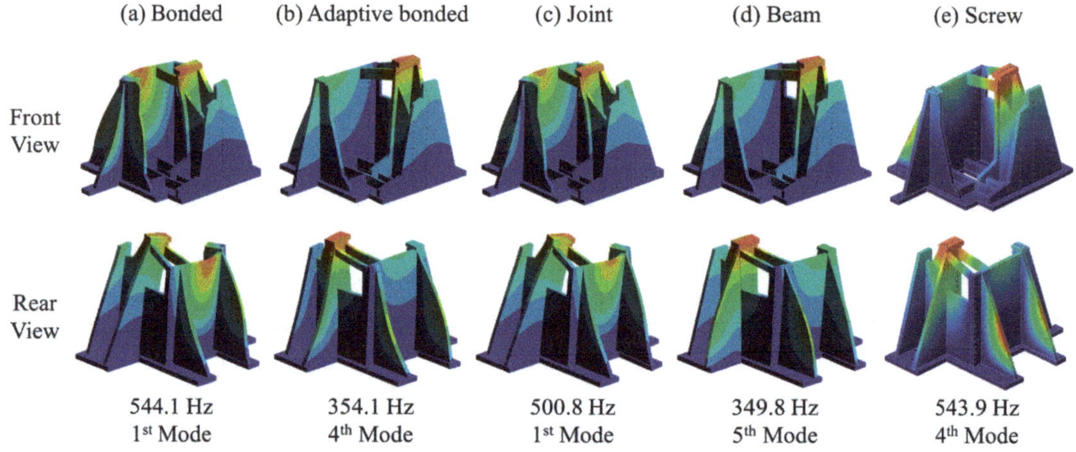

Figure 9. Natural frequencies and mode shapes based on different bolt connection methods: Top Head mode.

Table 2 summarizes the preprocessing time, CP Time, and memory usage for each bolt connection method. Preprocessing times were averaged over five trials conducted by experts to enhance the reliability of the analysis. The analysis efficiency and resource consumption varied significantly depending on the bolt connection method. Preprocessing time was particularly influenced by the number of bolts created and fastened. The adaptive bonded and joint methods required the longest times, each exceeding 6000 s (more than one hour), whereas the bonded, screw, and fixed bolt methods completed preprocessing in just 30 s. The beam method showed intermediate results, requiring approximately 3000 s. For CP Time, the adaptive bonded method took the longest, at 598 s, while the bonded method was the fastest, at 91.65 s. In terms of memory usage, the bonded method consumed the most memory at 37,393 MB, whereas the screw method was the most efficient, using only 25,590 MB.

Table 2. Comparison of preprocessing time, analysis time, and memory usage for different bolt connection methods.

	Bonded	Adaptive Bonded	Joint	Beam	Screw	Fixed Bolt
Average time required preprocessing [s]	30	6000	6000	3000	30	30
CP Time [s]	95	598	97	345	209	274
Memory usage [MB]	37,393	35,731	32,671	30,044	25,590	25,985

4. Discussion

This study analyzed the natural frequencies, mode shapes, and efficiency of various bolt implementation methods in a vibration test jig to propose the optimal connection method. To this end, the natural frequency and mode shape of the vibration test jig applied with six-bolt connection methods were compared to the natural frequency (493.2 Hz) derived from the Top Head in an actual vibration experiment. The joint connection method was found to produce the results most similar to the vibration test, with a difference of only 7.6 Hz, indicating high accuracy.

Through the results of this study, the discrepancy between vibration test results and bonded connection analysis was quantitatively assessed. The bonded connection method showed a natural frequency of 544.1 Hz, approximately 10.3% higher than the vibration test result (493.2 Hz). This is because the bonded contact exhibits greater stiffness than the bolt connections. Whereas bolt connections only connect specific points with gaps between joint surfaces, bonded connection assumes complete bonding between nodes, resulting in stronger coupling [35]. Therefore, although the bonded connection method is effective for simulating welded connections, it has limitations in representing the dynamic characteristics of bolt connections.

To determine whether the proposed bolt connection methods could simulate actual bolts, the differences in natural frequencies among the connection methods were examined. The two types of connections differed in structural stiffness, with bolt connection methods (except for fixed bolts) exhibiting relatively lower natural frequencies than bonded connections. This is because, unlike bonded connections, bolt connection methods involve smaller contact areas and account for nonlinear behavior in their coupling characteristics [36]. Additionally, the fixed bolt method exhibited a natural frequency of 881.9 Hz, which was 337.8 Hz higher than the natural frequency of 544.1 Hz for the bonded connection. This is because the fixed bolt method was based on Inspire's functionality, which treats the bolt head and endpoints of the threads as fixed points, restricting displacement and rotation degrees of freedom at both points. In other words, the fixed bolt method exhibits excessive rigidity, making it unsuitable for reproducing the dynamic characteristics of actual bolts. However, the fixed bolt method can be effectively utilized in designs requiring maximum structural coupling rigidity, such as heavy equipment, bridge structures, or stationary machinery.

A comparison of the modes among the analyzed bolt connection methods showed that the bonded connection method exhibited a global mode where the entire jig moved, while the bolt connection methods (except for the joint method) showed local modes with maximum deformation occurring in the ribs. This is because bolt connection methods have insufficient contact area and bonding strength, with nonlinear behavior influencing vibration characteristics [37–39]. On the other hand, the joint method, similar to the bonded connection, exhibited a global mode where the entire jig moved. This is because the joint method, based on ANSYS's joint functionality, rigidly connects two surfaces or points, providing high contact stiffness. As a result, relatively high contact stiffness in the rib connections occurs, effectively dispersing vibrations in local areas. Therefore, the joint method provides higher contact stiffness than other bolt implementation methods, significantly impacting structural stiffness and vibration behavior.

This study successfully identified a reliable bolt connection method for vibration test jigs through computational modal analysis and experimental validation, providing an approach for accurate dynamic behavior prediction. The modal analysis evaluated the dynamic behavior of the structure based on natural frequencies and mode shapes. Among the methods, the joint method indicated the most similar results, with approximately a 1.5% difference from the experimental values. These results are consistent with bolt connections for the dynamic analyses of an automotive engine [22], where a similar joint method showed a high correlation between the natural frequency of the computational analysis results and experimental results, further validating the reliability of this study. Furthermore, incorporating dynamic stress into this study's methodology could lead to a more refined design of structural dynamic behavior. A study on FE refinement for accurate dynamic stress [40] showed that considering dynamic stress in the evaluation process not only improved the structure's fatigue performance but also enhanced the reliability of the design under dynamic conditions. Thus, this approach could provide dynamic characteristic-based designs with greater practical applicability and improved overall performance.

In terms of analysis efficiency, notable differences between the connection methods were observed. The adaptive bonded method required the longest analysis time at 598 s, followed by the beam method at 345 s, while the joint method required a relatively short analysis time of 97 s. The screw and fixed bolt methods had analysis times ranging from approximately 200 to 300 s. The bonded method required the shortest analysis time at 91.65 s, but exhibited significant discrepancies from the test results, limiting its ability to reproduce the dynamic characteristics of actual structures. During preprocessing, the joint and adaptive bonded methods required approximately 2 min per bolt to generate line bodies and assign contact conditions, and the beam method also required significant time for assigning contact conditions, resulting in low preprocessing efficiency. By contrast, the screw and fixed bolt methods significantly reduced preprocessing time by utilizing Inspire's Fastener functionality, which automates preprocessing. As a result, while the joint method provided high analysis accuracy, it required substantial time for preprocessing, making it the second most time-consuming method overall. The adaptive bonded and beam methods also required significant time for both preprocessing and analysis, whereas the screw and fixed bolt methods demonstrated their potential as efficient alternatives in the early stages of design owing to their simplified preprocessing and short analysis times.

This study has some limitations. First, the analysis did not explicitly consider bolt preload and contact friction coefficients, indicating that the subtle effects that may occur under actual bolt fastening conditions were not fully captured. Second, the experiment was limited to excitation in the x-axis direction only, restricting a comprehensive evaluation of the dynamic behavior of the structure under various vibration directions. Third, the study was based on a single jig design case, limiting the generalizability of the findings to other structural designs. Nevertheless, this study demonstrated the utility of quantitatively comparing natural frequencies and mode shapes for various bolt implementation methods, providing valuable insights for the design and analysis of vibration test jigs.

5. Conclusions

This research systematically compared and analyzed the natural frequencies and mode shapes associated with different bolt implementation methods in the design of vibration test jigs. The bonded condition failed to adequately reproduce the dynamic characteristics of bolt connections, resulting in significant discrepancies between the analysis and actual vibration test results. However, the joint method exhibited the smallest difference in natural frequency compared to the test results, making it a reliable option for design. This suggests that the joint method can relatively accurately represent the contact conditions and stiffness of actual bolts, thus demonstrating that the choice of bolt connection method has a critical influence on the accuracy and efficiency of design and analysis. Although there were some differences between the analysis models and experiments, along with limitations in achieving complete consistency across software platforms, this study provides practical guidelines for selecting connection methods to consider during the early design stage. We believe that future studies incorporating diverse loading conditions and additional design cases will help in establishing more generalized criteria.

Author Contributions: Conceptualization, J.H.B. and J.J.K.; methodology, J.H.B., J.H.K., and J.B.B.; software, J.H.B. and J.J.K.; validation, J.H.B., J.H.K., and J.J.K.; formal analysis, J.H.B. and J.J.K.; investigation, J.H.B. and J.B.B.; resources, J.J.K.; data curation, J.H.B.; writing—original draft preparation, J.H.B. and J.B.B.; writing—review and editing, J.J.K.; visualization, J.H.B. and J.B.B.; supervision, J.J.K.; project administration, J.J.K. All authors have read and agreed to the published version of the manuscript.

Funding: This research was supported by the Korea Institute for the Advancement of Technology (KIAT) grant funded by the Korea Government (MOTIE) (P0021513, Industrial Innovation Infrastructure Establishment Project). Also, following are results of a study on the "Convergence and Open Sharing System" Project, supported by the Ministry of Education and National Research Foundation of Korea.

Institutional Review Board Statement: Not applicable.

Informed Consent Statement: Not applicable.

Data Availability Statement: The raw data supporting the conclusions of this article will be made available by the authors on request.

Conflicts of Interest: Author Ji Hye Kang was employed by the company SL Corporation. The remaining authors declare that the re-search was conducted in the absence of any commercial or financial relationships that could be construed as a potential conflict of interest.

Appendix A

Figure A1 shows an image of one rib in the finite element model of the jig with varying mesh sizes. To achieve a balance between efficiency and reliability, the mesh sizes were varied throughout the analysis. Detailed information for mesh sizes ranging from 8 mm to 4 mm is provided in Table A1. Additionally, Table A1 presents the number of nodes and elements, as well as the mesh quality, for each mesh size. The mesh quality was compared using the aspect ratio, which shows convergent values at a mesh size of 5 mm. Based on this evaluation, the mesh size for the jig model was set to 5 mm.

Figure A1. FE models of ribs according to mesh size.

Table A1. Number of nodes and elements and mesh quality according to mesh size.

Mesh Size [mm]	8	7	6	5	4
Number of Nodes	608,064	867,557	1,015,244	1,094,136	1,425,119
Number of Elements	343,811	496,721	582,436	627,725	820,739
Aspect Ratio	2.69	2.50	2.43	2.31	2.29
Jacobian Ratio	0.97	0.96	0.97	0.97	0.97
1st Frequency [Hz]	544.9	543.5	543.7	543.8	543.8

Appendix B

Table A2. Effect of Bolt Mesh Size on the First Natural Frequency.

Mesh Size [mm]	7	6	5	4	3	2
1st Frequency [Hz]	500.14	500.14	500.12	500.12	499.97	499.96

References

1. Oh, S.-H.; Kim, H.-J.; Park, K.-S.; Kim, J.-D. Evaluation of Static Displacement Based on Ambient Vibration for Bridge Safety Management. *Sensors* **2024**, *24*, 6557. [CrossRef]
2. Fan, W.; Qiao, P. Vibration-Based Damage Identification Methods: A Review and Comparative Study. *Struct. Health Monit.* **2011**, *10*, 83–111. [CrossRef]
3. Daouk, S.; Louf, F.; Cluzel, C.; Dorival, O.; Champaney, L.; Audebert, S. Study of the Dynamic Behavior of a Bolted Joint under Heavy Loadings. *J. Sound Vib.* **2017**, *392*, 307–324. [CrossRef]
4. Hearn, G.; Testa, R.B. Modal Analysis for Damage Detection in Structures. *J. Struct. Eng.* **1991**, *117*, 3042–3063. [CrossRef]
5. Martiny, P.; Lani, F.; Kinloch, A.J.; Pardoen, T. A Multiscale Parametric Study of Mode i Fracture in Metal-to-Metal Low-Toughness Adhesive Joints. *Int. J. Fract.* **2012**, *173*, 105–133. [CrossRef]
6. Hooper, J.M.; Marco, J. Defining a Representative Vibration Durability Test for Electric Vehicle (EV) Rechargeable Energy Storage Systems (RESS). *World Electr. Veh. J.* **2016**, *8*, 327–338. [CrossRef]
7. Moon, S.I.; Cho, I.J.; Yoon, D. Fatigue Life Evaluation of Mechanical Components Using Vibration Fatigue Analysis Technique. *J. Mech. Sci. Technol.* **2011**, *25*, 631–637. [CrossRef]
8. Česnik, M.; Slavić, J.; Boltežar, M. Uninterrupted and Accelerated Vibrational Fatigue Testing with Simultaneous Monitoring of the Natural Frequency and Damping. *J. Sound Vib.* **2012**, *331*, 5370–5382. [CrossRef]
9. Craig, O.; Picavea, J.; Gameros, A.; Axinte, D.; Lowth, S. Conformable Fixture Systems with Flexure Pins for Improved Workpiece Damping. *J. Manuf. Process.* **2020**, *50*, 638–652. [CrossRef]
10. Hooper, J.M.; Marco, J.; Chouchelamane, G.H.; Chevalier, J.S.; Williams, D. Multi-Axis Vibration Durability Testing of Lithium Ion 18650 NCA Cylindrical Cells. *J. Energy Storage* **2018**, *15*, 103–123. [CrossRef]
11. Sah, S.M.; Thomsen, J.J.; Brøns, M.; Fidlin, A.; Tcherniak, D. Estimating Bolt Tightness Using Transverse Natural Frequencies. *J. Sound Vib.* **2018**, *431*, 137–149. [CrossRef]
12. Liu, X.; Zhang, Y.; Yang, J. Non-Uniform Elastic Constraint Modeling and Stiffness Identification for the Bolted Thin-Walled Cylindrical Shell. *Thin-Walled Struct.* **2024**, *203*, 112233. [CrossRef]
13. Wang, H.; Xue, P.; Zhang, Y.; Jiang, L.; Wang, S. Research on the Influence of Disc–Drum Connection Bolt Preloading Rotor Assembly Modal Characteristics and Diagnosis Technology. *Machines* **2024**, *12*, 83. [CrossRef]
14. Gaul, L.; Nitsche, R. The Role of Friction in Mechanical Joints. *Appl. Mech. Rev.* **2001**, *54*, 93–106. [CrossRef]
15. Paroissien, E.; Lachaud, F.; Schwartz, S.; Da Veiga, A.; Barrière, P. Simplified Stress Analysis of Hybrid (Bolted/Bonded) Joints. *Int. J. Adhes. Adhes.* **2017**, *77*, 183–197. [CrossRef]
16. Ikbal, M.K.H.; Nguyen, D.V.; Kim, S.; Kim, D. Seismic Evaluation of Different Types of Electrical Cabinets in Nuclear Power Plants Considering Coupling Effects: Experimental and Numerical Study. *Nucl. Eng. Technol.* **2023**, *55*, 3472–3484. [CrossRef]
17. Pirdayr, A.; Mohammadi, M.; Javad Kazemzadeh-Parsi, M.; Rajabi, M. Self-Loosening Effects on Vibration Characteristics of Plates with Bolted Joints: An Experimental and Finite Element Analysis. *Meas. J. Int. Meas. Confed.* **2021**, *185*, 109922. [CrossRef]
18. Li, Y.; Long, T.; Luo, Z.; Wen, C.; Zhu, Z.; Jin, L.; Li, B. Numerical and Experimental Investigations on Dynamic Behaviors of a Bolted Joint Rotor System with Pedestal Looseness. *J. Sound Vib.* **2024**, *571*, 118036. [CrossRef]
19. Barbieri, N.; de Sant'Anna Vitor Barbieri, L.; Legat, J.T.; de Lima, K.F.; de Sant'Anna Vitor Barbieri, G. Bridge Damage Analysis Using Operational Modal Analysis. *J. Vib. Eng. Technol.* **2024**, *12*, 7799–7814. [CrossRef]
20. Wang, Y.; Liu, Y.; Xiong, F.; Zheng, C.; Ge, Q.; Bian, Y. Shaking Table Test of a Full-Scale Lightweight Bolt-Connected Concrete Sandwich Wall Panel Structure: Overview and Seismic Analyses. *J. Build. Eng.* **2024**, *97*, 110773. [CrossRef]

21. Paroissien, E.; Sartor, M.; Huet, J.; Lachaud, F. Analytical Two-Dimensional Model of a Hybrid (Bolted/Bonded) Single-Lap Joint. *J. Aircr.* **2007**, *44*, 573–582. [CrossRef]
22. Giannella, V.; Sepe, R.; Citarella, R.; Armentani, E. FEM Modelling Approaches of Bolt Connections for the Dynamic Analyses of an Automotive Engine. *Appl. Sci.* **2021**, *11*, 4343. [CrossRef]
23. Kim, J.; Yoon, J.-C.; Kang, B.-S. Finite Element Analysis and Modeling of Structure with Bolted Joints. *Appl. Math. Model.* **2007**, *31*, 895–911. [CrossRef]
24. Shamoto, E.; Hashimoto, Y.; Shinagawa, M.; Sencer, B. Analytical Prediction of Contact Stiffness and Friction Damping in Bolted Connection. *CIRP Ann.* **2014**, *63*, 353–356. [CrossRef]
25. Bhashyam, G.R. ANSYS Mechanical—A Powerful Nonlinear Simulation Tool. 2002. Available online: https://www.academia.edu/11917187/ANSYS_Mechanical_A_Powerful_Nonlinear_Simulation_Tool (accessed on 1 January 2025).
26. Wang, Y. Lightweight Steering Equipment Based on Prestressed Modal Analysis. *J. Vibroeng.* **2024**, *26*, 370–382. [CrossRef]
27. Brand, M.J.; Schuster, S.F.; Bach, T.; Fleder, E.; Stelz, M.; Gläser, S.; Müller, J.; Sextl, G.; Jossen, A. Effects of Vibrations and Shocks on Lithium-Ion Cells. *J. Power Sources* **2015**, *288*, 62–69. [CrossRef]
28. French, R.M.; Handy, R.; Cooper, H.L. A Comparison of Simultaneous and Sequential Single-Axis Durability Testing. *Exp. Tech.* **2006**, *30*, 32–37. [CrossRef]
29. Brownjohn, J.M.W.; Magalhaes, F.; Caetano, E.; Cunha, A. Ambient Vibration Re-Testing and Operational Modal Analysis of the Humber Bridge. *Eng. Struct.* **2010**, *32*, 2003–2018. [CrossRef]
30. Hao, J.; Zhu, X.; Li, J. Structural Damage Detection for Spatial Frame Structures with Semi-Rigid Joints Using Multiple Set Wireless Measurements. *J. Vib. Control.* **2024**, 1–15. [CrossRef]
31. Tang, Q.; Li, C.; Wen, B. Vibration Behaviors of Cylindrical Shell with Bolted Joint Effects: An Experimental Study. *Adv. Mech. Eng.* **2024**, *16*, 1–9. [CrossRef]
32. Hassiotis, S.; Jeong, G.D. Assessment of Structural Damage from Natural Frequency Measurements. *Comput. Struct.* **1993**, *49*, 679–691. [CrossRef]
33. Xu, G.Y.; Zhu, W.D.; Emory, B.H. Experimental and Numerical Investigation of Structural Damage Detection Using Changes in Natural Frequencies. *J. Vib. Acoust.* **2007**, *129*, 686–700. [CrossRef]
34. Lu, Z.R.; Law, S.S. Features of Dynamic Response Sensitivity and Its Application in Damage Detection. *J. Sound Vib.* **2007**, *303*, 305–329. [CrossRef]
35. Chowdhury, N.M.; Chiu, W.K.; Wang, J.; Chang, P. Experimental and Finite Element Studies of Bolted, Bonded and Hybrid Step Lap Joints of Thick Carbon Fibre/Epoxy Panels Used in Aircraft Structures. *Compos. Part B Eng.* **2016**, *100*, 68–77. [CrossRef]
36. Overend, M.; Nhamoinesu, S.; Watson, J. Structural Performance of Bolted Connections and Adhesively Bonded Joints in Glass Structures. *J. Struct. Eng.* **2013**, *139*, 1–12. [CrossRef]
37. Lacayo, R.; Pesaresi, L.; Groß, J.; Fochler, D.; Armand, J.; Salles, L.; Schwingshackl, C.; Allen, M.; Brake, M. Nonlinear Modeling of Structures with Bolted Joints: A Comparison of Two Approaches Based on a Time-Domain and Frequency-Domain Solver. *Mech. Syst. Signal Process.* **2019**, *114*, 413–438. [CrossRef]
38. Jamia, N.; Jalali, H.; Taghipour, J.; Friswell, M.I.; Haddad Khodaparast, H. An Equivalent Model of a Nonlinear Bolted Flange Joint. *Mech. Syst. Signal Process.* **2021**, *153*, 107507. [CrossRef]
39. Huda, F.; Kajiwara, I.; Hosoya, N.; Kawamura, S. Bolt Loosening Analysis and Diagnosis by Non-Contact Laser Excitation Vibration Tests. *Mech. Syst. Signal Process.* **2013**, *40*, 589–604. [CrossRef]
40. Zhou, Y. Local Finite Element Refinement for Accurate Dynamic Stress via Modal Information Only. *AIAA J.* **2020**, *58*, 3593–3606. [CrossRef]

applied sciences

MDPI

Article

Topology Optimization and Testing of Connecting Rod Based on Static and Dynamic Analyses

Mahalingam Nainaragaram Ramasamy [1,*], Aleš Slíva [1], Prasath Govindaraj [2] and Akash Nag [2]

1 Institute of Transport, Faculty of Mechanical Engineering, VSB-Technical University of Ostrava, 17. Listo-padu 15/2172, 70800 Ostrava, Czech Republic

2 Department of Mechanical Technology, Faculty of Mechanical Engineering, VŠB–Technical University of Ostrava, 70800 Ostrava, Czech Republic; prasath.govindaraj.st@vsb.cz (P.G.)

* Correspondence: mahalingam.nainaragaram.ramasamy.st@vsb.cz

Abstract: This research article outlines our aim to perform topology optimization (TO) by reducing the mass of the connecting rod of an internal combustion engine based on static structural and dynamic analyses. The basic components of an internal combustion engine like the connecting rods, pistons, crankshaft, and cylinder liners were designed using Autodesk Inventor Professional 2025. Using topology optimization, we aimed to achieve lesser maximum von Mises stress during static structural analysis and maintain a factor of safety (FOS) above 2.5 during rigid body dynamics. A force of 64,500 N was applied at the small end of the connecting rod while the big end was fixed. Topology optimization was carried out using ANSYS Discovery software at various percentages on a trial-and-error basis to determine better topology with lesser maximum von Mises stress. Target reduction was set to 4%, and as a result, 5.66% mass reduction from the original design and 6.25% reduced maximum von Mises stress was achieved. Later, transient analysis was carried out to evaluate the irregular motion loads and moments acting on the connecting rod at 1000 rpm. The results showed that the FOS remained above 2.5. Finally, the optimized connecting rod was simulated and verified for longevity using Goodman fatigue life analysis.

Keywords: topology optimization; connecting rod; finite element method; multibody dynamics; factor of safety; fatigue life analysis

1. Introduction

The connecting rod is the most crucial and highly dynamic component of internal combustion engines, which helps in converting the reciprocating motion of the piston into the rotational motion of the crankshaft [1]. The connecting rod consists of a small end, shank, and big end, where the small end is connected to the piston using a piston pin, and the big end is attached to the crankshaft [2]. Irregular motion loads and moments due to crankshaft rotation combined with the combustion forces from the piston acting on the connecting rod play a vital role in optimizing this structure without affecting its longevity. Frictional energy loss plays an important role in mechanical efficiency and accounts for between 10% and 15% of the energy produced by internal combustion engines. Components of the engine and drivetrain, such as the piston rings, valve guides, cam followers, fuel injector pistons, shims, transmission gears, and front seals, are to varying degrees affected by friction. The selection of appropriate materials and lubricants and the proper design and size of an engine's functional parts are not only necessary to minimize friction and wear on operational components but also to reduce friction and wear on engine components. The proper size of an engine can reduce fuel consumption by about 9% [3]. Finite element method (FEM) strength analysis is used to determine the cause of damage, evaluate structural integrity, and optimize design to prevent failure by simulating the distribution of stress and material behavior under different loading conditions [4]. In the past few years, computer-aided design (CAD) tools like SolidWorks 2024 and Autodesk Inventor 2025, and FEM tools like ABAQUS 2023HF2, NASTRAN 2023.4, etc., have evolved, and the TO process plays a major role in product improvement. The storage and reuse of the data from the CAD model for future improvement, adaptation, or use in different applications is widely implemented by many product developers [5].

CAD models are simulated with structural analysis to understand their deformation, displacement, etc. Structural analysis is a method used to study behavior under static loading conditions depending on material properties, mesh, and boundary conditions [6]. According to the Kisaru Shveli method, the FOS at the connecting rod body should be above 1.5, and at the big end and small end, it should be above 2.5 [7]. Dynamic analysis is a method used to evaluate the kinematics of assembly by studying motion while a system is kept rigid from deformation [8]. The application of the TO process has a large scope depending on the requirement of structural enhancements in fields like the automotive industry [9], robotics [10], aerospace [11], and industrial machinery [12]. There are several methods used to achieve TO, such as the density-based method, level-set method, evolutionary structural optimization, etc. [13]. The TO process is a computational method used to reduce the mass and the maximum von Mises stress of a defined boundary condition, while ensuring structural integrity, optimizing material distribution, and maintaining a suitable factor of safety for reliable performance under applied loads [13,14].

However, despite its application, limited research has addressed the TO of highly dynamic components like connecting rods, crankshafts, etc. The main objective of this research was to perform TO on a connecting rod using the finite element method and multibody dynamics by reducing the mass and maximum von Mises stress during static structural analysis. Unlike conventional topology optimization studies that focused exclusively on mass reduction, this research simultaneously optimized the distribution of stress and ensured structural integrity by maintaining a safety factor (FOS) above 2.5. The current study also incorporated static structural analysis, dynamic analysis, and fatigue life analysis (Goodman criteria) to fully evaluate the long-term durability of optimizing designs. In addition to the static TO, this study validated an optimal design under dynamic load and fatigue life analysis, ensuring that the connecting rod was stable under an irregular load and moment of motion.

Furthermore, material restrictions were introduced to ensure that optimized designs could be manufactured using traditional production methods such as forging and casting. These material savings can contribute to fuel efficiency and improved engine performance. An optimized design was evaluated by transitional analysis and fatigue under realistic load conditions to ensure that it remained structurally sound under operational stress. The results demonstrate that, when combined with rigorous structural validation, topology optimization can significantly improve the performance of components while reducing unnecessary material use.

This paper is structured in the following way: Section 1 (Introduction) explains the problem and objective of this research along with a description of the proposed solution; Section 2 (Materials and Methodology) describes the properties of the chosen material and the design of the IC engine components; Section 3 (Computational Analysis) describes the methodologies used; Section 4 (Results) includes a discussion and comparison of the results; Section 5 (Conclusion) explains the outcome of the research and future scope of this work; and the References section contains the bibliography of this work.

2. Materials and Methodology

2.1. Design and Material Selection

All the IC engine components including the connecting rod were designed using Autodesk Inventor Professional 2025. Since the objective of this study was to optimize the topology of the connecting rod, all other components were set as rigid bodies during the transient structural analysis and suppressed during the TO process.

The connecting rod was mostly designed with an I-Section for normal IC engines and an H-Section for race engines, and the mass of the connecting rod should be somewhere between 0.4 kg and 1 kg. The connecting rod ratio (λ) is defined by the relationship between the crank radius (r_c) and the length of the connecting rod from the center of the small end and the big end (l_{cr}). Typically for passenger car engines, the connecting rod ratio (λ) should be between 0.28 and 0.32 [15].

$$\lambda = \frac{r_c}{l_{cr}} \quad (1)$$

$$\lambda = \frac{45}{153} = 0.2941 \quad (2)$$

Here the connecting rod ratio (λ) was around 0.30 and it satisfied this condition [15]. The major dimensions of the original connecting rod geometry are shown in Figure 1.

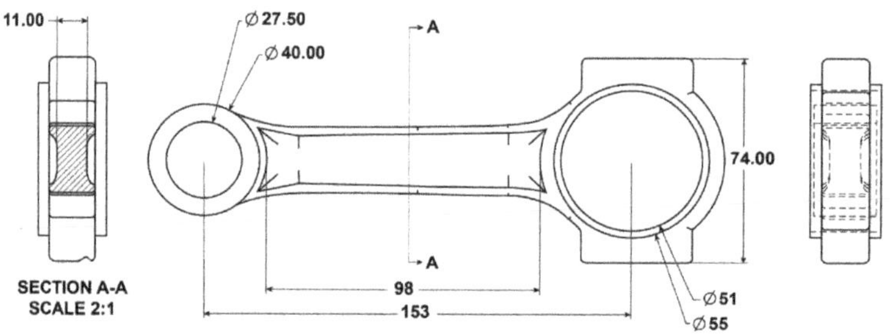

Figure 1. Major dimensions of the connecting rod's initial design.

2.1.1. Material Selection

The chosen connecting rod was made from 36MnVS4 and is used in the second generation of fracture-split material. This is the latest technology used while manufacturing connecting rods nowadays; in this method, the connecting rod is manufactured as a single component and then it is split into two halves at the crank end [16]. Based on the selected material, the mass of the initial connecting rod was 654 g. The major composition of this low-alloy steel is manganese and vanadium which improve its durability to withstand tensile, compressive, and buckling loads. Other materials like carbon, silicon, sulfur, phosphorus, and chromium can also be found in small quantities [17]. The properties of 36MnVS4 steel are shown in Table 1.

Table 1. Properties of 36MnVS4 [16].

Yield Strength	Tensile Strength	Density	Modulus of Elasticity	Poisson's Ratio
800 MPa	1000 MPa	7.85 g/cm^3	210 GPa	0.29

2.1.2. Design of Engine Components

Other components of the engine were designed to yield a low speed and high torque, based on the following assumed parameters: 4-cylinder inline engine, cylinder bore of

81 mm, and cylinder stroke length of 82.25 mm, which could run at a maximum speed of 1000 rpm in Table 2.

Table 2. Assumed engine parameters for the initial design of the connecting rod.

Engine Type	Cylinder Bore	Stroke Length	Engine Speed
Inline 4-cylinder	81 mm	82.25 mm	1000 rpm (Max)

2.1.3. Assembly of Engine Components

Since this assembly was designed only for motion study and kinematic analysis, the components in this IC engine assembly were minimized to four major components, the crankshaft, piston, connecting rod, and cylinder liner, avoiding small components like piston rings, bearings, etc. The precision of the other parts was not given priority, and the parts were designed to be close to fit and used only to evaluate the kinetics of unknown reaction forces and moments acting on the connecting rod. The assembly was made with the help of revolute joints, fixed joints, and translational joints. Figure 2 shows the assembly of the cylinder liner, piston, and crankshaft from Autodesk Inventor 2025.

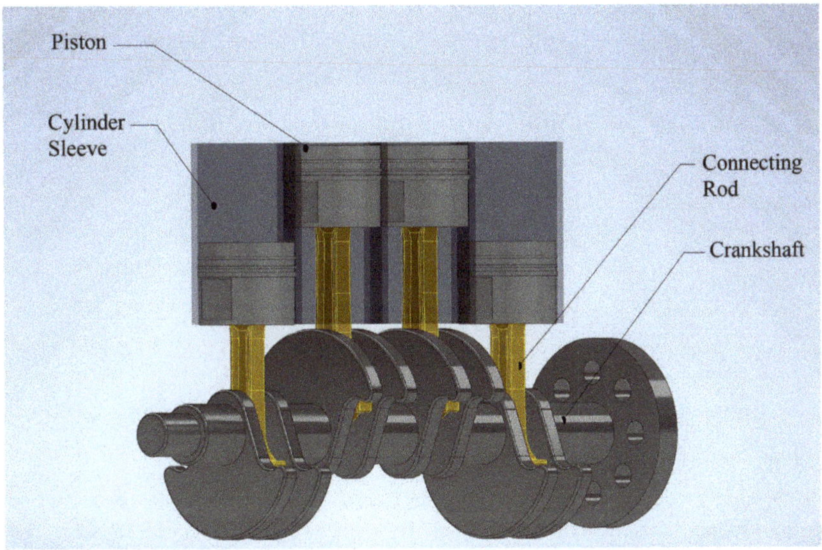

Figure 2. Virtual assembly of the crankshaft, connecting rod, piston, and cylinder liner.

The assembly design was exported from Autodesk Inventor Professional 2025 as a STEP file to have compatibility with ANSYS Discovery and Workbench.

2.2. Methodology

In this study, the finite element method, multibody dynamics, and topology optimization were carried out using ANSYS 2024 R2 and ANSYS Discovery 2024 R2 software. The TO process was conducted for various percentages on a trial-and-error basis to determine the minimum von Mises stress and the results were verified for their longevity during dynamic conditions. The whole TO process from the connecting rod, based on static structural analysis, to the validation of the topology in multibody dynamics is shown below in Figure 3.

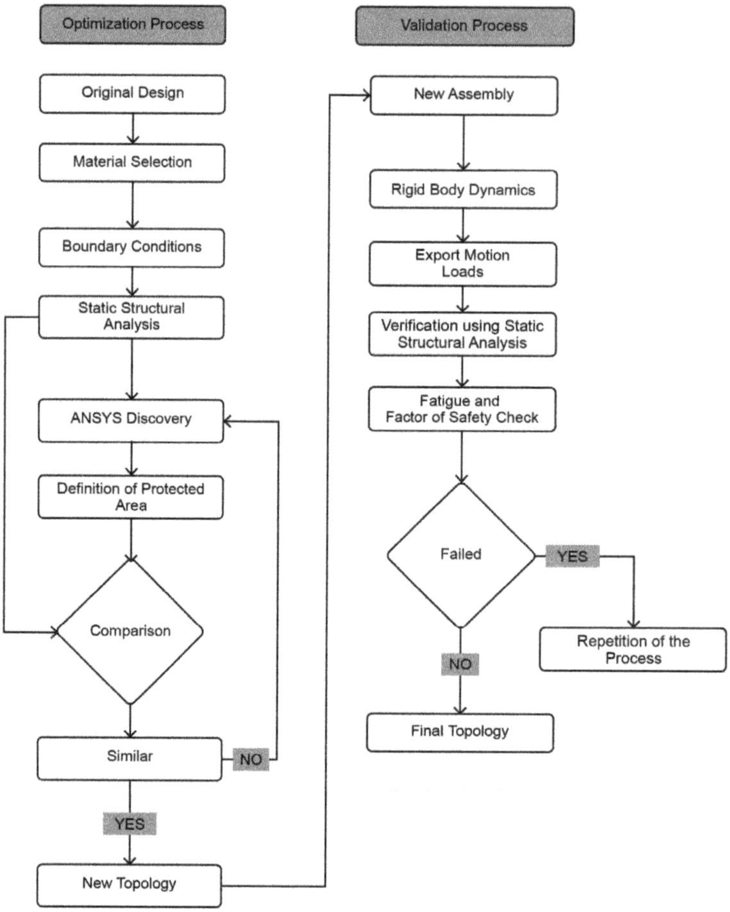

Figure 3. Topology optimization and validation process workflow.

3. Computational Analysis

In this study, we conducted various analyses to evaluate the performance of the connection rods, including static and dynamic analyses. These analyses helped to verify the structural integrity, load capacity, and possible failure modes under different operating conditions.

3.1. Static Structural Analysis

Static structural analysis was performed in the ANSYS workbench, and engineering data were added to the material library for the material 36MnVS4. Then, the initial connecting rod's CAD model was imported into the ANSYS environment. Figure 4 shows the boundary conditions and mesh properties of the static structural analysis used in the initial design of the connecting rod. Mesh properties are important in structural analysis, as higher mesh resolution leads to more accurate simulation results [18]. As a boundary condition, the big end of the connecting rod was fixed, and a load of 64,500 N was applied to the small end as shown in Figure 4a. The mesh was then generated, as shown in Figure 4b.

Calculation of force acting on the piston due to combustion (f_C).

For gasoline engines, their maximum peak pressure is between 70 bar and 130 bar [19].

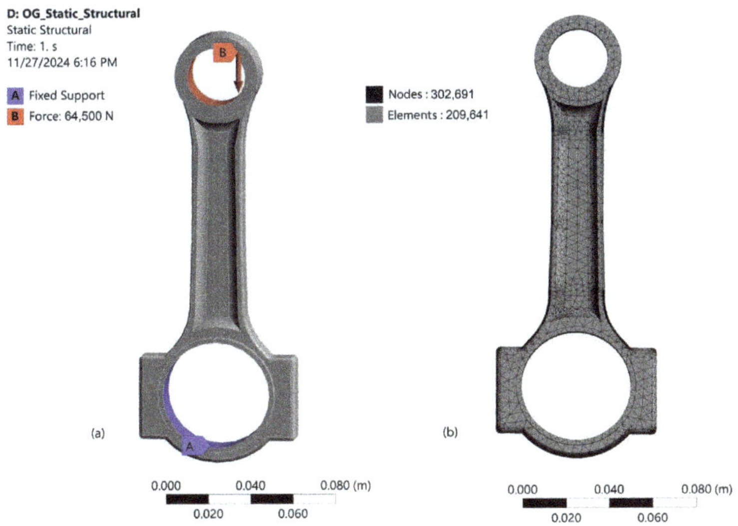

Figure 4. (**a**) Boundary conditions applied during the static structural analysis in the initial design. (**b**) Mesh properties used during the static structural analysis in the initial design.

Assuming the peak pressure (P) as 125 bar for the gasoline engine, the area of the piston (A) was determined using a piston diameter (d_p) of 81 mm in the following equation:

$$A = \pi \times \left(\frac{d_p}{2}\right)^2 \quad (3)$$

The area of the piston (A) was therefore 0.005153 m^2.

$$fc = P \times A \quad (4)$$

$$fc = 64{,}412.5 \approx 64{,}500\ \text{N} \quad (5)$$

The force acting on the piston due to combustion (fc) was 64,500 N.

As a boundary condition, the big end of the connecting rod was set as fixed, and the combustion force of 64,500 N was applied directly on the piston and carried to the small end of the connecting rod. When the combustion force was applied on the small end of the connecting rod, a buckling load acted on the shank part of the connecting rod as a result, causing it to experience compressive stress, which may have led to bending or deformation. Later, meshing was performed and 302,691 nodes with 209,641 elements were generated for a resolution of 5, slow transition, and fine span angle center as shown in Figure 4. Once the mesh was generated, the loading conditions were applied based on the peak pressure.

Based on the boundary conditions, the obtained maximum von Mises stress was 5.8556×10^8 Pa and the maximum von Mises strain was 0.0028.

3.2. Topology Optimization

ANSYS Discovery software uses the Solid Isotropic Material with Penalization (SIMP) method. During the iteration process, the ANSYS Discovery solver adjusts the material density, (ρ) from 0 (ρ) to 1 within the permissible area outside the protected depth. Herein, if the material density (ρ) is equal to 1, it represents full material density, and if the (ρ) is equal to 0, it represents zero material density [20]. There are several options for the TO process in ANSYS Discovery, i.e., optimizing for the strength or response to free vibration, optimizing for a specified frequency, optimizing to minimize stress, and optimizing to minimize volume [21].

$$(\rho) = \rho^p \times K \quad (6)$$

where K denotes the real stiffness matrix of an element, and p denotes the factor of penalization [14,22].

The following targets were outlined for the TO process:

- To reduce the mass at least by 3%;
- To reduce the maximum von Mises stress and von Mises strain;
- To define the protected depths at both ends of the connecting rod;
- To maintain major dimensions;
- To obtain a different cross-section than the I-Section;
- To keep the FOS above 2.5;
- To manufacture the component in both additive manufacturing and conventional forging.

In this research, minimizing stress was set as the objective of the TO process. While setting the boundary conditions, the same conditions were used as in the static structural analysis, and by defining the protected depths of 0.012 m on both ends of the connecting rod, the TO process could be carried out as shown in Figure 5 [23]. The TO process was carried out at various target percentages from 3% to 5% to reduce the mass of the existing design [24].

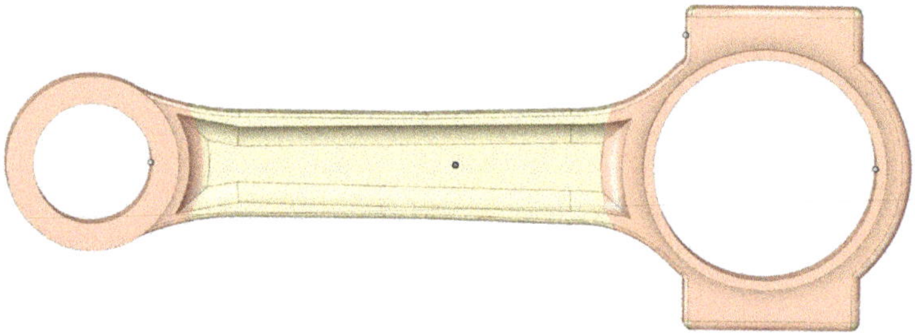

Figure 5. Protected depth of 0.012 m on both ends of the connecting rod.

Two of the important factors influencing the optimization process were the quality of the mesh and the number of elements. Once the topology was achieved, facets were created for that percentage of reduced mass, and we were required to clean up the facet design using the repair facet tool and subD hybrid option as shown in Figure 6, before exporting it as an STL or STEP file [25].

The goal of the TO in this study was to minimize mass while ensuring structural integrity by controlling the von Mises stress.

$$\min_{\rho} M(\rho) = \int \rho \, d\Omega \quad (7)$$

where ρ is the material density function, $M(\rho)$ is the total mass of the connecting rod, and Ω denotes the design domain.

$$\min_{\rho} \sigma_{max}(\rho) \quad (8)$$

where σ_{max} is the maximum von Mises stress in the structure [26,27].

The connecting rod was subject to the following constraints.

The optimized design had to ensure that von Mises stress did not exceed an allowable limit:

$$\sigma_{max}(\rho) \leq \sigma_{allow} \quad (9)$$

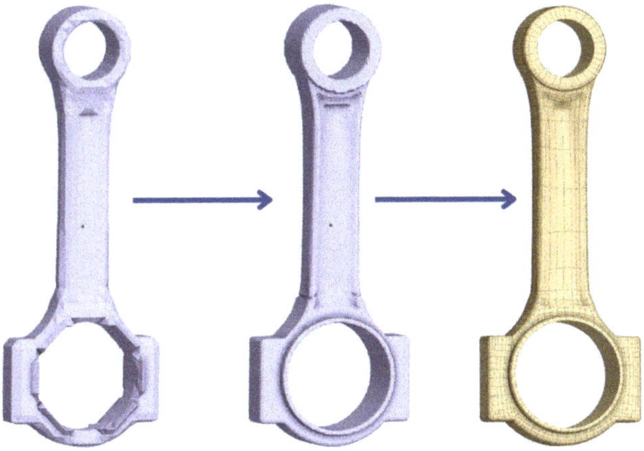

Figure 6. The sequence of the connecting rod TO from initial facets to the final subD hybrid conversion.

To ensure structural safety, the minimum FOS had to be maintained as the following:

$$FOS = \frac{\sigma_{yield}}{\sigma_{max}} \geq 2.5 \quad (10)$$

As a boundary condition, the big end of the connecting rod was fixed, and force was applied to the small end of the connecting rod [28].

After solving the topology at various percentages on a trial-and-error basis, the results targeted at 4% mass reduction had a mass reduction of 5.66% with minimum von Mises stress compared to the targeted values of 3% and 5%. Figure 7a shows the original shape of the connecting rod, and Figure 7b shows the final optimized topology of the connecting rod with a 5.66% mass reduction and initial design.

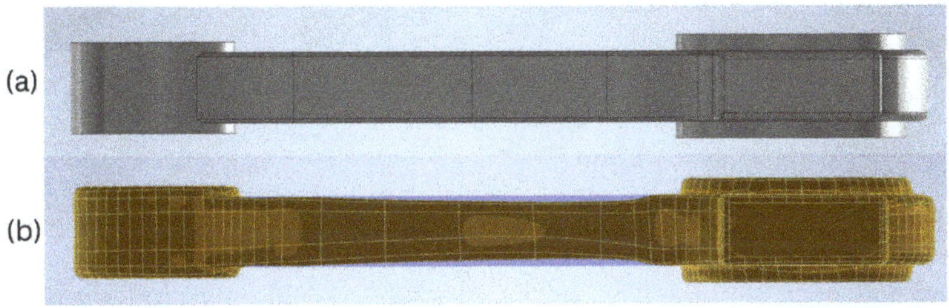

Figure 7. (a) Original shape of the connecting rod. (b) 5.66% reduced mass from the original shape of the connecting rod.

The obtained topology had a 5.66% reduction in mass from 654 g to 618 g.

3.3. Dynamic Analysis

The objective of the dynamic analysis was to determine the unknown reaction forces and moments acting on the connecting rod while the engine was running at the speed of 1000 rpm. During the dynamic analysis, all the bodies were defined as rigid bodies and assembled using transitional, fixed, and revolute joints. The rotational velocity (ω) was determined based on the defined maximum rpm. For the speed of 1000 rpm, the rotational velocity is defined as

$$\omega = \frac{RPM \times 2\pi}{60} = 104.72 \text{ rad/s} \quad (11)$$

Once the connections were defined according to Table 3, the maximum unknown reaction forces acting on the connecting rod were probed using a joint probe in the solution section. Since all bodies were set as rigid, there was no meshing involved while determining the dynamic motion loads.

Table 3. Connections used in the engine assembly.

Connections Between the Parts	Joints Used	Degrees of Freedom
Cylinder liners to ground	Fixed joint	0 DOF
Crankshaft to ground	Fixed revolute joint	1 DOF
Cylinder to piston	Transitional joint	4 (x-axis)
Piston to connecting rod	Revolute joint	(2) × 4 DOFs
Connecting rod to crankshaft	Revolute joint	(2) × 4 DOFs

While constructing the assembly, it was important to calculate the total degree of freedom (DOF), by assuming the required kinematic motions for each part individually [29]. This assembly had 13 degrees of freedom, and it is explained below as shown in Table 3.

A rotational velocity (ω) of 104.72 rad/s was applied on the crankshaft as shown in Figure 8a. Once the motion study was completed, the maximum reaction loads as shown in Figure 8b,c were retrieved from the solution and exported as a text file to evaluate the reaction forces acting on the connecting rod in a separated static structural analysis.

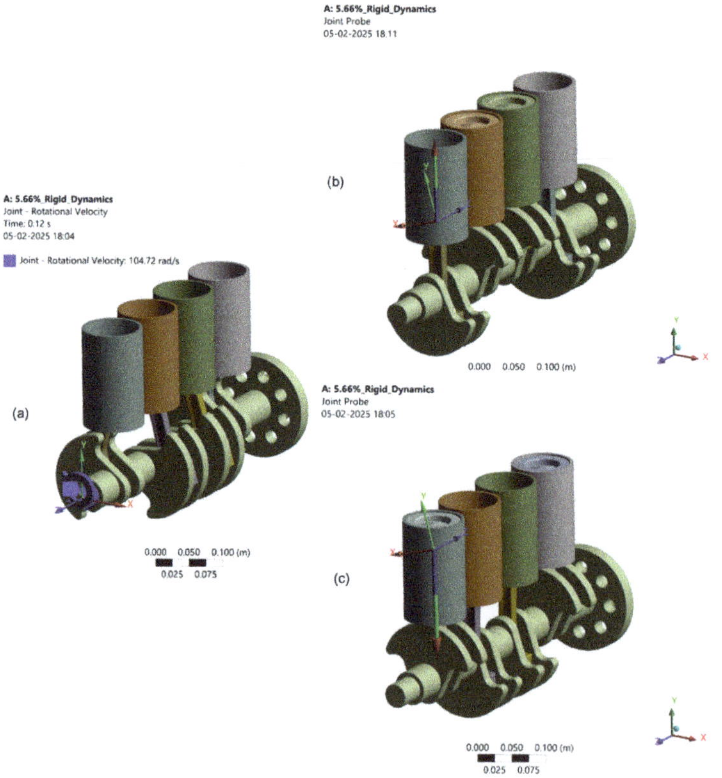

Figure 8. (**a**) Boundary conditions—A rotational velocity of 104.72 rad/s was applied to the crankshaft. (**b**,**c**) Maximum unknown reaction forces extracted using a joint probe while the piston was moving toward the top dead center and the bottom dead center.

From this study, the engine assembly had a good balance between the counterweight of the crankshaft and the newly obtained topology of the connecting rod, which is due to a minimum weight reduction and optimized topology. When the weight reduction was targeted at more than 7%, there was a noticeable imbalance and increase in vibration due to the connecting rod weight ratio and the counterweight of the crankshaft.

Motion Loads

Motion loads are the loads that are obtained from the motion of a mechanical system with certain allowable DOFs, such as inertial loads due to the mass of moving parts, impact loads due to sudden collisions, and dynamic loads from cyclic motion or vibrations, etc. In this study, motion loads were important since the engine underwent rapid tensile and compressive forces during cyclic motion along with inertial forces due to the motion of the crankshaft and connecting rod.

The obtained motion load from the rigid body dynamics was applied to a new static structural study, where the connecting rod was subjected to two reaction forces, and two tensile forces were applied simultaneously along with two reaction moments while moving toward the top dead center (TDC), as shown in Figure 9a,b.

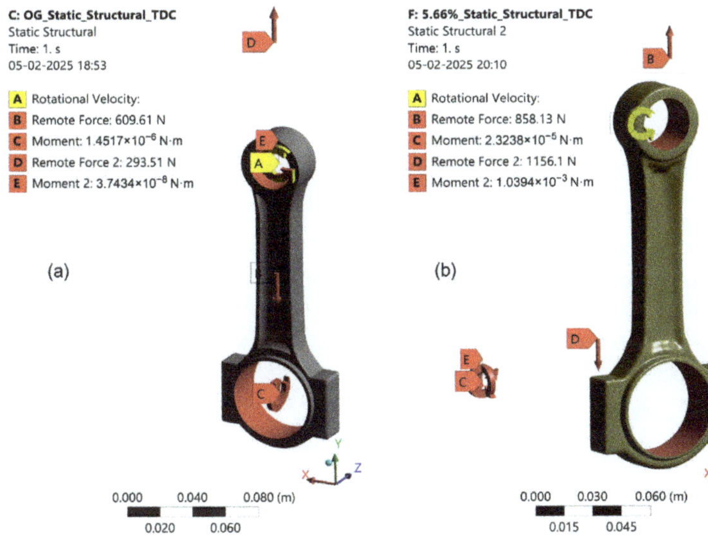

Figure 9. (**a**) Maximum unknown reaction forces applied during static structural analysis on the initial design while at the TDC. (**b**) Maximum unknown reaction forces applied during static structural analysis on the optimized topology with a mass reduction of 5.66% while at the TDC.

These reaction force values as shown in Figure 9a explains When the connecting rod moved toward the TDC, there were remote forces of 609.61 N and 293.51 N acting as tensile forces, along with moments of 1.4517×10^{-6} Nm and 3.7434×10^{-8} Nm acting on the initial design of the connecting rod.

Figure 9b shows that when the connecting rod moved toward the TDC, there were remote forces of 858.13 N and 1156.1 N acting as tensile forces, along with moments of 2.3238×10^{-5} Nm and 1.0394×10^{-8} Nm acting on the optimized topology of the connecting rod with a mass reduction of 5.66%.

The forces obtained from the rigid body dynamics were applied to a new static structural study where the connecting rod was exposed to two compression forces at the same time as two reaction moments while it moved toward the bottom dead center (BDC), as shown in Figure 10a,b.

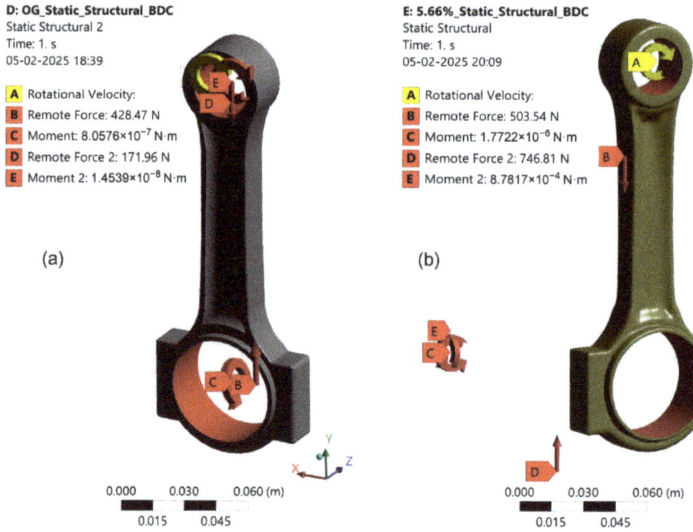

Figure 10. (**a**) Maximum unknown reaction forces applied during static structural analysis on the initial design while at the BDC. (**b**) Maximum unknown reaction forces applied during static structural analysis on the optimized topology with a mass reduction of 5.66% while at the BDC.

These reaction force values as shown in Figure 10a explains When the connecting rod moved toward the BDC, there were remote forces of 428.47 N and 171.96 N acting as compressive forces, along with moments of 8.0576×10^{-7} Nm and 1.4539×10^{-8} Nm acting on the initial design of the connecting rod.

Figure 10b shows that when the connecting rod moved toward the BDC, there were remote forces of 503.54 N and 746.81 N acting as compressive forces along with moments of 1.7722×10^{-6} Nm and 8.7817×10^{-4} Nm acting on the optimized topology of the connecting rod with a mass reduction of 5.66%.

These reaction force values show how the engine balance varied between the initial design of the connecting and optimized topology for 5.66% weight reduction. When the mass of the connecting rod was reduced by more than 7%, that may have caused a total imbalance between the counterweight of the crankshaft and the connecting rod, making the engine unstable by causing high vibrations. To avoid this problem, the ratio between the crankshaft counterweight and connecting rod should always be maintained.

4. Results

This section has two parts, wherein the first part explains the results obtained directly from the static structural analysis, and the second part discusses the verification of the optimized topology from the dynamic analysis.

4.1. Results from Static Structural Analysis

The maximum von Mises stress obtained from the original topology for the applied load of 65 KN was 5.8556×10^8 Pa, as shown in Figure 11a, while the minimum von Mises stress was 630.8 Pa, with an average von Mises stress of 6.3227×10^7 Pa. Once the topology was optimized for the targeted 4%, the mass was originally reduced by up to 5.66% from 654 g to 618 g. The maximum von Mises stress obtained for the same load of 65 KN was

5.5006×10^8 Pa, as shown in Figure 11b, and the minimum von Mises stress was 291.7 Pa, with an average von Mises stress of 1.3862×10^8 Pa.

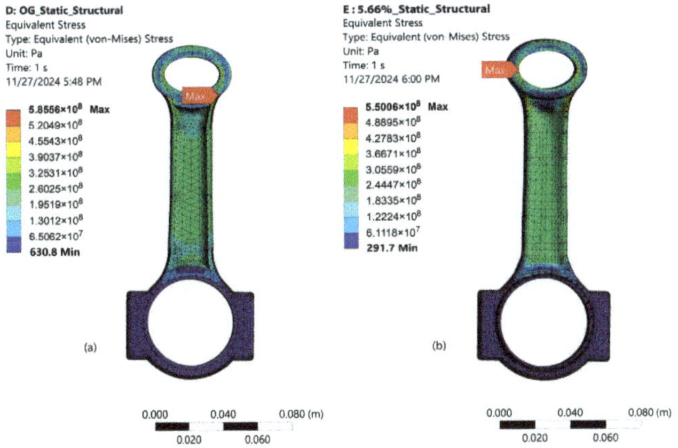

Figure 11. (**a**) Maximum von Mises stress acting on the initial design. (**b**) Maximum von Mises stress acting on the 5.66% mass-reduced optimized topology.

4.2. Results from Motion Loads

The dynamic study was carried out to extract the motion loads to evaluate the safety factor of the connecting rod. As discussed in Section 3, the connecting rod was subjected to simultaneously tensile and compressive loads, and only the maximum reaction forces acted while the connecting rod moving toward the TDC and BDC was evaluated. Goodman fatigue life analysis was used to evaluate the FOS and life of the connecting rod and verify that the optimized topology did not fail under the defined conditions. Based on the given reaction forces, the connecting rod had the lowest safety factors of 2.8476 and 2.6182 at the crank end while moving toward the TDC and BDC, respectively, as shown in Figure 12a,b. Given that the FOS needed to be more than 2.5 for the connecting rod, the results satisfied this condition since the obtained FOS values were above 2.5 even in the weaker nodes.

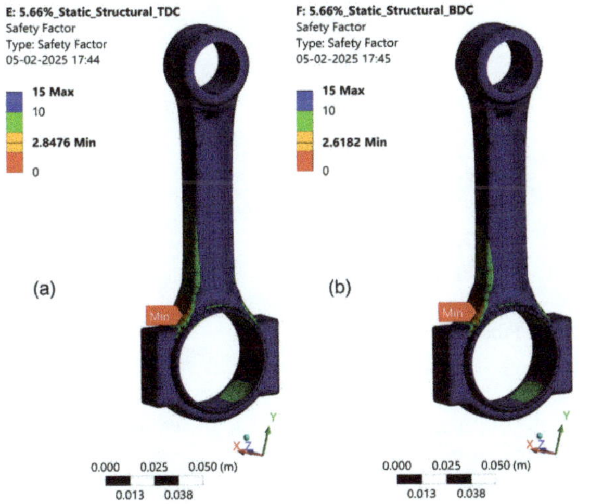

Figure 12. (**a**) Safety factor after applying the motion loads on the 5.66% mass-reduced optimized topology at the TDC. (**b**) Safety factor after applying the motion loads on the 5.66% mass-reduced optimized topology at the BDC.

Goodman fatigue life analysis was used since the chosen material was alloy-steel and to predict the lifecycle of the connecting rod when it was subjected to varying stress [30].

Contributions of this research

1. The TO process of the connecting rod showed the possibilities of optimizing similar highly dynamic components.
2. An improvement in the existing design was achieved by reducing the mass by 5.66%.
3. The static structural results were validated by the dynamic analysis.
4. The obtained optimized topology was designed in such a way to make it possible to be produced by both conventional manufacturing methods like forging and additive manufacturing.
5. The combination of static structural analysis for the design and rigid body dynamics for the verification of the optimized design can be repeated and applied to similar studies.

The results clearly show that the connecting rod did not fail under the given motion load conditions while moving toward the TDC and BDC, even at 1 million cycles, as shown in Figure 13a,b.

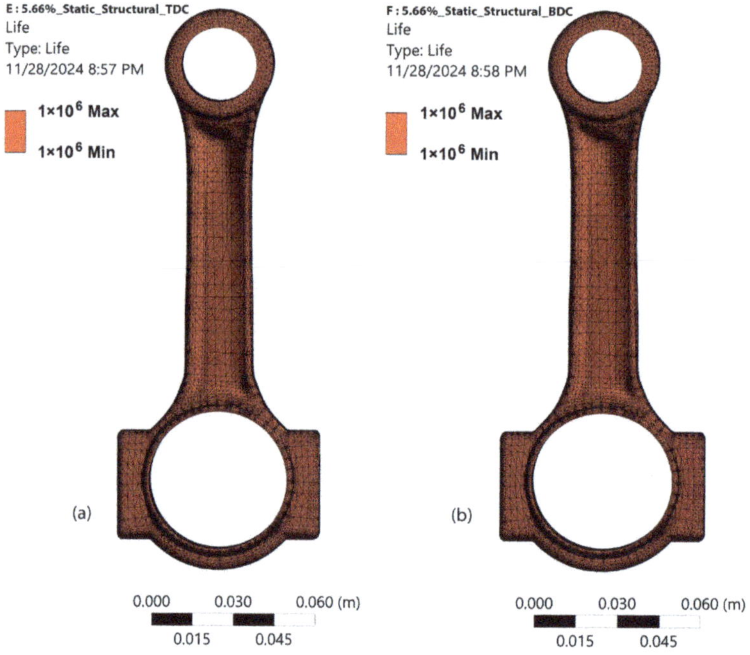

Figure 13. (**a**) Life of the connecting rod after applying the motion loads on the 5.66% mass-reduced optimized topology at the TDC. (**b**) Life of the connecting rod after applying the motion loads on the 5.66% mass-reduced optimized topology at the BDC.

5. Conclusions

This research aimed to optimize the topology of a connecting rod by reducing the mass of the component and minimizing the maximum von Mises stress. The optimization process included various procedures including static structural analysis and dynamic analysis for the validation of the FOS and the life of the connecting rod based on Goodman fatigue analysis. A motion study was carried out as part of the validation, where the maximum unknown reaction forces and moments acting on the connecting rod were evaluated while the connecting rod moved toward the TDC and BDC. The main obtained results were the following:

- The topology of the connecting rod was optimized to 5.66% from 654 g to 618 g.
- The maximum von Mises stress was reduced from 5.8556×10^8 Pa to 5.5006×10^8 Pa during static structural analysis.
- The FOS was maintained above 2.6 during the dynamic analysis of the connecting rod.

As part of this study, it was also found that the excess reduction in the mass of the connecting rod had a direct effect on the stability of the engine. Hence, the weight ratio between the connecting rod and the counterweight of the crankshaft should be maintained.

A possible continuation of this research could be conducted by manufacturing the topology-optimized connecting rod using additive manufacturing or by other conventional methods and experimenting with different materials to improve the properties of the connecting rod. Later, the new connecting rod could be tested using the digital image correlation method (DIC) to validate the simulation results with the actual results.

Author Contributions: Conceptualization, M.N.R.; methodology, M.N.R.; software, M.N.R. and P.G.; validation, A.S. and A.N.; formal analysis, M.N.R.; investigation, M.N.R. and P.G.; data curation, M.N.R.; writing—original draft, M.N.R. and P.G.; writing—review and editing, M.N.R., A.S., and A.N.; visualization, M.N.R.; supervision, A.S.; project administration, M.N.R.; funding acquisition, A.S. All authors have read and agreed to the published version of the manuscript.

Funding: This work was supported in part by the Student Grant Competition project "Research, Development, and Innovation in the Field of Transport and Logistics 2025" financed by the Ministry of Education, Youth, and Sports and Faculty of Mechanical Engineering, Vysoka skola banska-Technical University of Ostrava (VSB-TUO), under SP2025/065; by the Materials and Technologies for Sustainable Development within the Jan Amos Komensky Operational Program financed by the European Union (EU) and the State Budget of Czech Republic under Grant No. CZ.02.01.01/00/22_008/0004631; and by the EU through REFRESH—Research Excellence For Region Sustainability and High-Tech Industries via the Operational Programme Just Transition Project No. CZ.10.03.01/00/22_003/0000048.

Institutional Review Board Statement: Not applicable.

Informed Consent Statement: Not applicable.

Data Availability Statement: The data presented in this paper can be made available on request from the corresponding author.

Acknowledgments: All authors have read and agreed to the published version of the manuscript. The authors would like to thank the Faculty of Mechanical Engineering and the Ph.D. Academy of Vysoka skola banska-Technical University of Ostrava (VSB-TUO) for their support during the computational study using Autodesk Inventor Professional 2025, ANSYS 2024 R2 and guidance with academic writing.

Conflicts of Interest: The authors declare no conflicts of interest.

Abbreviations

The following abbreviations are used in this manuscript:

TO	Topology optimization
FOS	Factor of safety
FEM	Finite element method
RPM	Revolutions per minute
CAD	Computer-aided design
IC	Internal combustion
STEP	Standard for the Exchange of Product

SIMP	Solid Isotropic Material with Penalization
STL	Stereolithography
DOF	Degree of freedom
TDC	Top dead center
BDC	Bottom dead center
DIC	Digital image correlation

References

1. Heywood, J.B. *Internal Combustion Engine Fundamentals*; McGraw-Hill: New York, NY, USA, 1988.
2. *J604D_197906*; Engine Terminology and Nomenclature-General. SAE: Warrendale, PA, USA, 1979.
3. Kocaaslan, O. Numerical investigation on pulse detonation engine performance under different equivalence ratio. *Adv. Eng. Lett.* **2024**, *3*, 1–12. [CrossRef]
4. Decker, Ž.; Rudzinskas, V.; Drozd, K.; Caban, J.; Tretjakovas, J.; Nieoczym, A.; Matijošius, J. Analysis of the Vehicle Chassis Axle Fractures. *Materials* **2023**, *16*, 806. [CrossRef] [PubMed]
5. Camba, J.D.; Contero, M.; Company, P. Parametric CAD Modeling: An Analysis of Strategies for Design Reusability. *Comput.-Aided Des.* **2016**, *74*, 18–31. [CrossRef]
6. Rao, S.S. *The Finite Element Method in Engineering*; Butterworth-Heinemann: Oxford, UK, 2017.
7. Gao, W.; Wang, G.; Zhu, J.; Fan, Z.; Li, X.; Wu, W. Structural Optimization Design and Strength Test Research of Connecting Rod Assembly of High-Power Low-Speed Diesel Engine. *Machines* **2022**, *10*, 815. [CrossRef]
8. Shabana, A.A. *Dynamics of Multibody Systems*; Cambridge University Press: Cambridge, UK, 2013; ISBN 9781107042650.
9. Caban, J.; Nieoczym, A.; Matijošius, J.; Kilikevičius, A.; Drozd, K. Analysis of the construction of the car trailer frame in terms of changing the assembly technology. *Sci. J. Silesian Univ. Technol. Ser. Transp.* **2024**, *124*, 47–61. [CrossRef]
10. Wang, R.; Zhang, X.; Zhu, B.; Zhang, H.; Chen, B.; Wang, H. Topology Optimization of a Cable-Driven Soft Robotic Gripper. *Struct. Multidiscip. Optim.* **2020**, *62*, 2749–2763. [CrossRef]
11. Okorie, O.; Perveen, A.; Talamona, D.; Kostas, K. Topology Optimization of an Aerospace Bracket: Numerical and Experimental Investigation. *Appl. Sci.* **2023**, *13*, 13218. [CrossRef]
12. Mirzendehdel, A.M.; Behandish, M.; Nelaturi, S. Topology Optimization with Accessibility Constraint for Multi-Axis Machining. *Comput.-Aided Des.* **2020**, *122*, 102825. [CrossRef]
13. Bendsøe, M.P.; Sigmund, O. *Topology Optimization: Theory, Methods and Applications*; Springer: Berlin/Heidelberg, Germany, 2004; ISBN 978-3-662-05086-6.
14. Mikulikova, A.; Mesicek, J.; Karger, J.; Hajnys, J.; Ma, Q.-P.; Sliva, A.; Smiraus, J.; Srnicek, D.; Cienciala, S.; Pagac, M. Topology Optimization of the Clutch Lever Manufactured by Additive Manufacturing. *Materials* **2023**, *16*, 3510. [CrossRef] [PubMed]
15. Richard van Basshuysen, F.S. *Internal Combustion Engine Handbook: R-345: Basics, Components, Systems and Perspectives*; SAE International: Warrendale, PA, USA, 2004.
16. Kou, S.; Shi, Z.; Song, W. Fracture-Splitting Processing Performance Study and Comparison of the C70S6 and 36MnVS4 Connecting Rods. *SAE Int. J. Engines* **2018**, *11*, 463–474. [CrossRef]
17. Yan, J.; Chen, L.; Zhang, C.; Miao, H.; Wang, S.; Wu, G.; Gao, J.; Wu, H.; Zhao, H. Influence of Sulfur on the Splitting Fracture and Machining Performance of Microalloyed Medium-Carbon Steel 36MnVS4 Connecting Rods. *J. Mater. Eng. Perform.* **2024**, *33*, 12125–12132. [CrossRef]
18. Shaari, M.S.; Rahman, M.M.; Noor, M.M.; Kadirgama, K.; Amirruddin, A.K. Design of Connecting Rod of Internal Combustion Engine: A Topology Optimization Approach. In Proceedings of the National Conference in Mechanical Engineering Research and Postgraduate Studies, Kuantan, Pahang, Malaysia, 3–4 December 2010; pp. 155–166.
19. Hamm, T.; Rebbert, M.; Ecker, H.-J.; Grafen, M. Cylinder Head Design for High Peak Firing Pressures. *SAE Int. J. Engines* **2008**, *1*, 746–755. [CrossRef]
20. Bendsøe, M.P.; Kikuchi, N. Generating Optimal Topologies in Structural Design Using a Homogenization Method. *Comput. Methods Appl. Mech. Eng.* **1988**, *71*, 197–224. [CrossRef]
21. Topology Optimization. Available online: https://ansyshelp.ansys.com/public/account/secured?returnurl=/Views/Secured/corp/v242/en/discovery/UDA/user_manual/physics/structural/topics/c_topo_intro.html (accessed on 3 October 2024).
22. User Guide, Altair OptiStruct. Available online: https://2021.help.altair.com/2021/hwsolvers/os/topics/solvers/os/topology_opt_design_variables_r.htm (accessed on 10 October 2024).
23. Setting the Protected Depth. Available online: https://ansyshelp.ansys.com/public/account/secured?returnurl=//////Views/Secured/corp/v242/en/discovery/UDA/user_manual/physics/structural/topics/t_topo_protect_distance.html (accessed on 24 September 2024).

24. Topology Optimization Goals and Tool Options. Available online: https://ansyshelp.ansys.com/public/account/secured?returnurl=//////Views/Secured/corp/v242/en/discovery/UDA/user_manual/physics/structural/topics/r_structural_topo_tools.html (accessed on 12 September 2024).
25. Creating the SubD Hybrid Model. Available online: https://ansyshelp.ansys.com/public/account/secured?returnurl=/Views/Secured/corp/v242/en/discovery/UDA/user_manual/modeling/facets/topics/t_autoskin_create_subd_hybrid.html?q=subD%20hybrid (accessed on 12 October 2024).
26. Duysinx, P.; Bendsøe, M.P. Topology Optimization of Continuum Structures with Local Stress Constraints. *Int. J. Numer. Methods Eng.* **1998**, *43*, 1453–1478. [CrossRef]
27. Senhora, F.V.; Giraldo-Londoño, O.; Menezes, I.F.M.; Paulino, G.H. Topology Optimization with Local Stress Constraints: A Stress Aggregation-Free Approach. *Struct. Multidiscip. Optim.* **2020**, *62*, 1639–1668. [CrossRef]
28. Shigley, J.E.; Mischke, C.R. *Standard Handbook of Machine Design*; McGraw-Hil: New York, NY, USA, 1986.
29. Xu, L.; Li, Q.; Zhang, N.; Chen, Q. Mobility, Kinematic Analysis, and Dimensional Optimization of New Three-Degrees-of-Freedom Parallel Manipulator With Actuation Redundancy. *J. Mech. Robot.* **2017**, *9*, 041008. [CrossRef]
30. Rice, R.C. *SAE Fatigue Design Handbook AE-22*; SAE International: Warrendale, PA, USA, 1997.

applied sciences

MDPI

Article

Sketch-Guided Topology Optimization with Enhanced Diversity for Innovative Structural Design

Siyu Zhu [1,†], Jie Hu [1,*,†], Jin Qi [1,*,†], Lingyu Wang [2], Jing Guo [1], Jin Ma [3] and Guoniu Zhu [4]

1 School of Mechanical Engineering, Shanghai Jiaotong University, Shanghai 200240, China
2 School of Design, Shanghai Jiaotong University, Shanghai 200240, China
3 School of Electric Engineering, Shanghai Jiaotong University, Shanghai 200240, China
4 Academy for Engineering and Technology, Fudan University, Shanghai 200433, China
* Correspondence: hujie@sjtu.edu.cn (J.H.); jinhuaqj@sjtu.edu.cn (J.Q.)
† These authors contributed equally to this work.

Abstract: Topology optimization (TO) is a powerful generative design tool for innovative structural design, capable of optimizing material distribution to generate structures with superior performance. However, current topology optimization algorithms mostly target a single objective and are highly dependent on the problem definition parameters, causing two critical issues: limited human controllability and solution diversity. These issues often lead to burdensome design iterations and insufficient design exploration. This paper proposes a multi-solution TO framework to address them. Human designers express their stylistic preferences for structures through sketches which are decomposed into stroke and closed-shape elements to flexibly guide each TO process. Sketch-based constraints are integrated with Fourier mapping-based length-scale control to enhance human controllability. Solution diversity is achieved by perturbing Fourier mapping frequencies and load conditions in the neural implicit TO framework. Adaptive parallel scale adjustment is incorporated to reduce the computational cost for design exploration. Using the structural design of a wheel spoke as a case study, the mechanical performance and diversity of the generated TO solutions as well as the effectiveness of human control are analyzed both qualitatively and quantitatively. The results reveal that the sketch-based constraints and length-scale control have distinct control effects on structural features and have different impacts on the mechanical performance and diversity, thereby enabling fine-grained and flexible human controllability to better balance conflicting objectives.

Keywords: topology optimization; structural design; innovative design; free-hand sketch; human control; design diversity

1. Introduction

Topology optimization (TO) is a powerful tool for innovative structural design. By optimizing material distribution within a given design domain, it can generate conceptual designs that meet specified objectives and constraints, offering novel structures with superior engineering performance [1–5]. It has been widely applied in tasks like lightweight optimization and also shows great potential in other fields including heat transfer [6,7], fluid dynamics [8], photonics [9,10], etc.

An important feature of TO is that it eliminates the need for parameterization, avoiding the limitations of manually defined design parameters, and offering a broad, unconstrained design space. As such, it is an ideal generative design algorithm for the early conceptual design phase. Currently, TO algorithms for structural design can generate highly complex and optimized structural features, surpassing parametric design solutions in both mechanical strength and material efficiency [11]. However, they have two key limitations that hinder their broader application: a lack of efficient human control over the generated structural features and a limited diversity of solutions [12,13]. Since design exploration involves multiple complex and hard-to-quantify factors, a practical approach is to search for a diverse set of good solutions rather than a single optimized one, with cross-domain experts analyzing, filtering, and determining the final optimal design through iterations. In the collaboration between design experts and TO algorithms, enabling a human to directly and precisely control some structural features, such as making the minimal length scale of structures equal to a preset value, is crucial to reducing iterative processes. Meanwhile, solution diversity is key to exploring high-quality and novel designs.

Recently, studies focusing on enhancing human controllability and solution diversity for TO algorithms have attracted increasing attention. To address the first challenge, some studies [14–18] have focused on controlling the length scale of generated geometric features, as the length scale is critical for issues like local stress concentration, structural complexity, and manufacturability. Other approaches focus on guiding the TO process using specific geometric patterns, such as periodic cells [19–22], image content [23], or a reference design [24]. Among these, sketch-based TO guidance methods [12,25] demonstrate efficient controllability and are particularly intuitive for designers accustomed to expressing ideas through sketches. However, existing studies only provide a single way to achieve human control, limiting the flexibility in managing the fine-grained features and hindering the achievement of proper balance among conflicting design factors. Furthermore, though human control inevitably impacts design exploration freedom and reduces solution diversity, there are no existing studies that have tried to quantify this effect or explored methods to address it.

To address the second challenge, existing studies typically enhance solution diversity by perturbing the load condition parameters or adjusting the multi-objective weights [12,26]. Some studies [13,26–28] also propose data-driven approaches to enhance diversity by learning implicit representations from existing TO datasets, followed by random sampling in the learned latent feature space. However, studies on TO solution diversity remain lacking, and to our knowledge, no studies have quantitatively analyzed TO solution diversity or compared different diversifying approaches. Existing qualitative results indicate insufficient diversity, and show that diversifying solutions by merely perturbing hyperparameters has a significant negative impact on the mechanical performances of structures.

To address the aforementioned challenges in enhancing collaboration between human designers and generative algorithms in innovative structural design, this paper proposes a novel multi-solution TO framework. This framework integrates two approaches to achieve human controllability: sketch-based constraints and length-scale control of geometric features. Free-hand sketches are decomposed into stroke elements and closed-shape elements, with two distinct constraints introduced for them accordingly: a stroke-based undirected structural boundary constraint and a closed-shape-based regional constraint. They are further integrated with length-scale control of geometric features to provide multifaceted control strategies. Each control strategy, with unique control characteristics for structural features, has a different impact on design quality and diversity, thereby enabling fine-grained and flexible human controllability. Additionally, the framework enhances solution diversity by introducing randomness through the perturbation of Fourier mapping frequencies and load condition parameters in the density-based neural implicit TO process. To reduce the computing cost caused by having multiple TO processes, this framework implements an adaptive parallel computing scale adjustment method. A case study of an innovative design of automobile wheel spokes is conducted, and comprehensive studies on human controllability and its impact on both design quality and diversity are carried out qualitatively and quantitatively.

The main contributions of this paper are summarized as follows:

1. Novel sketch-based constraints, including a stroke-based undirected structural boundary constraint and a closed-shape-based regional constraint, are proposed to enable sketch-based TO control.
2. Sketch-based constraints and length-scale control of geometric features are integrated, providing multifaceted control strategies to enable fine-grained and flexible control.
3. A diversifying method based on random sampling of Fourier mapping frequencies in a neural implicit TO process is proposed, and quantitative studies on different diversifying methods as well as the effects of human control strategies on design quality and diversity are conducted.
4. An adaptive parallel computing scale adjustment method is incorporated to improve the efficiency of multiple TO processes.

2. Related Works

2.1. Human Control in Topology Optimization

The main advantage of TO is that designers only need to specify the design objectives, constraints, and load conditions within the design domain, without the need to provide prior knowledge for structural design parameterization. Most existing TO algorithms, particularly the earlier developed ones [29–31], provide an optimal solution of structural topology for a TO problem according to the design factor of mechanical performance. However, as numerous factors need to be considered in engineering practice, the TO algorithms fail to produce satisfactory solutions directly, leading to laborious iterations between human designers and the algorithms [32,33]. This problem became more pronounced due to the immature post-processing techniques of TO solutions. Effective human control over TO solutions is the key to this problem, which is becoming a highlight topic in this research field.

Recent studies on this problem have focused on controlling the length scale of geometric features [14–18], as the length scale is very important not only for mechanical performance but also other factors like manufacturability. Traditional TO algorithms tend to generate excessively fine-grained structures, which can incur problems in addictive manufacturing process such as local collapse or difficulty in support material removal. Ha et al. [15] and Schiffer et al. [16] achieved interactive control of the feature sizes within a region of interest through a proximity-based weighting function and three-phase projection scheme, respectively. The latter further enabled control of multi-phase maximum and minimum solid feature sizes and/or void feature sizes. Li et al. [34] explicitly derived the numerical formulation of minimum structural feature size constraints for density-based TO algorithms, offering broad applicability and low sensitivity to algorithm parameters. He et al. [35] introduced a hole-filling strategy in the bidirectional evolutionary structural optimization (BESO) framework to control structural complexity by filling internal cavities and tunnels, respectively, with preset structures whose size parameters were adjustable. Chandrasekhar et al. [14] introduced a Fourier mapping layer in a density-based neural implicit TO framework to control the maximal and minimal geometric feature size. Lazarov et al. [36] provided a detailed review and comparative analysis of studies on length-scale control for density-based TO algorithms.

Research on human controllability also includes using specific geometric patterns [19–22]. Zhang et al. [19] used predefined texture patterns to guide the optimization, dividing the design domain into multiple groups, where the repetition of the control pattern within each group was managed independently. Meng et al. [20] achieved similar controllability by arranging predefined patterns to determine the design domain. Li et al. [21] derived specific geometric constraints for the internal regions of shell structures to make the generated structures match the predefined patterns. Zhang et al. [25] proposed using free-hand sketches for TO guidance by extracting multi-level morphological features from design images using convolutional neural networks and introducing a similarity constraint between the extracted morphological features of TO solutions and the sketch. Their proposed method generates structural designs that embody the high-level morphological semantics of the sketch. Zhang et al. [23] used aesthetic images as guidance, employing independent feature extraction neural modules for the aesthetic style and form content of the images, respectively, which were then used to apply separate constraints. Zhong et al. [37] proposed exploiting the text–image priors embedded in the CLIP model [38] to enable text-based control over the aesthetic styling of the generated structures. Li et al. [12] proposed combining drawing guidance with manual scoring to modify TO solutions interactively. The sensitivity numbers of BESO within the affected regions of drawn strokes were adjusted, while the sensitivity numbers could also be globally adjusted according to manual scoring feedback.

Overall, length-scale control of geometric features has been enabled by several existing methods with its effectiveness and reliability validated, while guidance with specific geometric patterns is relatively immature. A proper length-scale control method can be selected according to customized design requirements for a specific TO framework, such as a density-based solid isotropic material, with penalization (SIMP) or BESO algorithms. The key challenge here is to properly determine the length-scale parameters [14,16]. For guidance with specific geometric patterns, existing studies can be divided into two categories: leveraging domain knowledge, or reference designs based on similarity constraints. Methods of the first category are sensitive to the application scenario and highly dependent on expertise, which limits their usability. Methods of the second category encounter inevitable conflicts between the similarity constraint and other objectives like mechanical performance optimization. Current methods tackle this only by a single weight coefficient of the similarity constraint, which is insufficient to reach a proper balance.

To enhance the flexibility of human control over TO solutions and better balance the various design factors, this paper combines sketch-based constraints with length-scale control. The sketches are decomposed into stroke elements and closed-shape elements, with different constraint terms and penalty weights assigned, respectively, to enable fine-grained control of different structural features. As studies on TO solution diversity and the impact of human controllability on diversity are lacking, this paper fills this gap as well.

2.2. Deep Learning-Enhanced Topology Optimization

Deep learning (DL) [39] has shown great potential in learning latent features from large datasets, facilitating tasks like new content exploration in latent spaces and prediction in decision-making tasks. In the field of TO, integrating DL methods has emerged as a promising solution to some of the field's tough challenges. Such studies can be broadly categorized into three types: (1) developing black-box TO solver models that replace traditional iterative TO algorithms [27,40–43]; (2) fusing DL with traditional algorithms [13,44–46]; and (3) using DL models to replace the finite element analysis (FEA) process, thus accelerating the iterations [47–49].

Among the first type of study, Yu et al. [42] used an encoder–decoder framework based on convolutional neural networks (CNNs) to establish a mapping from boundary conditions and optimization settings to the solutions of traditional TO algorithms. They further integrated conditional generative adversarial networks (cGANs) to improve image resolution of the outputs. Nie et al. [40] proposed a cGAN-based TO solver model, and incorporated multi-physical field simulation results with the load and boundary conditions to improve the quality of the generated designs. Maze et al. [41] replaced cGANs with a diffusion model [50,51] as the TO solver, and introduced a surrogate model to ensure structural performance and manufacturability. Zheng et al. [43] employed a U-Net model [52] as a 3D TO solver. Yoo et al. [27] proposed learning low-dimensional latent space from existing TO designs with a convolutional auto-encoder, then generated diverse designs via a sampling strategy based on design of experiment (DOE) in latent space.

Another branch of studies focused on fusing DL methods with iterative TO algorithms. Chandrasekhar et al. [44] encoded the density function of the Solid Isotropic Material with Penalization (SIMP) method using neural networks (NNs), combining the sensitivity analysis through automatic back-propagation of the NNs and independent FEA module. Chen et al. [45] introduced the FEA results of the initial material distributions as conditional input to the neural-encoded density function. Their results showed improved structural performance and higher computing efficiency. Wang et al. [46] embedded TO within a cGAN framework with loss terms of structural compliance and volume constraints to guide the optimization. Jang et al. [13] used reinforcement learning in an iterative fusion pipeline to improve design diversity.

A significant challenge in TO is the computational cost associated with the FEA. To address this, Qian and Ye [48] proposed a dual-model artificial neural network for forward computation and sensitivity analysis of the FEA, with discrepancy loss between the predicted and true FEA results. Yin et al. [49] introduced physics-informed neural networks (PINNs) to simulate the forward and sensitivity analysis of the FEA instead. Li et al. [47] proposed using a structure-mapping neural network to refine coarse-grained TO solutions into fine-grained ones at each iteration, thereby enabling an FEA to be conducted on low-resolution grids.

Although DL-based TO solvers beat traditional iterative algorithms in inference efficiency and design diversity, the generated designs mostly fail in satisfying mechanical constraints and manufacturability, making the DL-based solvers invalid as direct solutions to design problems. Using DL-based methods to replace FEA is effective in accelerating iterations while having a small influence on final outputs. One of the obstacles to using data-driven methods is the difficulty in establishing large datasets. Despite current advancements in incorporating DL, traditional iterative TO algorithms continue to play a critical role in ensuring the reliability and quality of the final designs. Fusion of them to fully leverage their respective strengths has great potential for achieving more efficient, high-quality, and diverse topology optimization.

3. Method

Figure 1 shows the proposed multi-solution TO framework for solving the typical structural design problem of minimizing compliance under volume constraint. Multiple parallel TO processes for solving the same problem are employed with consistent algorithm hyper-parameters. In each process, a SIMP-based method with a neural network that implicitly encodes the density function over the continuous domain is implemented. The volume constraint and sketch-based constraints are transformed into penalty loss terms and combined with the compliance loss term through a weighted sum, serving as the total training loss for the neural-encoded density function. The neural input signals are Fourier frequency components of sampled grid points. The high- and low-frequency thresholds of the input signals determine the length scale of the spatial geometric features. The Fourier sampling frequencies and load force magnitudes are randomly perturbed across TO processes to produce diverse design solutions. The TO processes are synchronized, and have their solution set filtered by removing redundant solutions at regular optimization iteration intervals after the first several iterations, which adaptively adjusts the parallel scale. Filtering is according to the similarity metrics in terms of geometry and design objective values.

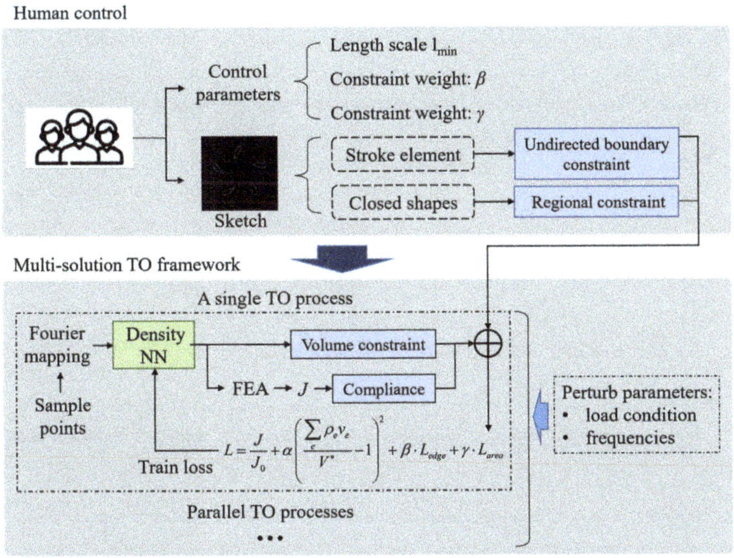

Figure 1. Overview of proposed multi-solution TO framework with human controllability.

The remainder of this section gives details of the proposed framework and the application of it in an innovative structural design scenario.

3.1. TO Process with Neural-Encoded Density Function

The mathematical formulation of solving the aforementioned design problem with the SIMP method is given in Equation (1). The optimizable variables of the design problem are the material density values of the discrete elements, $\rho = \{\rho_e\}$, where ρ_e is the material density of the e-th element within the value range [0, 1]. $\rho_e = 1$ represents that this element has solid material and $\rho_e = 0$ represents that this element is void. The design objective is to minimize the total structural compliance C.

For a specific design problem to be solved, the user-defined hyper-parameters include the applied force $\boldsymbol{F}$, the prescribed volume V^*, the element stiffness matrix for solid material element $\boldsymbol{K}_e^0$, the material's Young's modulus E_0, the element volume v_e, and the penalization exponent p. In Equation (1), the first constraint, $\boldsymbol{K}(\boldsymbol{\rho})\boldsymbol{u} = \boldsymbol{F}$, is the static equilibrium equation, where $\boldsymbol{K}(\cdot)$ is the finite element stiffness matrix as a function of the element densities $\boldsymbol{\rho}$ assembled from the element stiffness matrix $\boldsymbol{K}_e(\boldsymbol{\rho})$ computed by Equation (2). $\boldsymbol{u}$ is the displacement field computed from the static equilibrium equation. The second constraint represents the volume constraint.

$$\begin{aligned} \min_{\rho} \quad & C = \boldsymbol{F}^\top \boldsymbol{u} = \boldsymbol{u}^\top \boldsymbol{K}(\boldsymbol{\rho})\boldsymbol{u} \\ \text{subject to} \quad & \boldsymbol{K}(\boldsymbol{\rho})\boldsymbol{u} = \boldsymbol{F} \\ & \sum_e \rho_e v_e \leq V^* \end{aligned} \tag{1}$$

$$K(\rho) = [K_e(\rho)] = [K_e^0 E_0 \rho_e^p] \tag{2}$$

In the SIMP method, the element densities are continuous within the range [0, 1], and are pushed to 0 or 1 with increasing penalty value (typically initialized as $p = 1$ and gradually increased), thereby generating an approximately binary topology design solution. The element densities are optimized on the basis of an FEA of structural compliance conducted on fixed sample grids using domain-specific methods such as Optimality Criteria (OC) or Method of Moving Asymptotes (MMAs). This approach limits the resolution of the solution and requires an analytical sensitivity analysis. This study adopts the neural-encoded density function proposed by Chandrasekhar and Suresh [44], which leverages the high non-linearity of neural networks to represent a continuous density function over the design field in the SIMP method. This transforms the discrete element density optimization into continuous density function learning, alleviating the limitation of the solution resolution. This also allows different sample grids to be applied for various FEA objectives, for better balance between FEA accuracy and computation cost. Additionally, a sensitivity analysis can be automatically performed using the back-propagation of the neural network. The neural-encoded density function is denoted as $\rho_w(\cdot)$, where w is the neural weights.

3.2. Length-Scale Control

This study adopts the Fourier mapping method [14] to achieve structural feature length-scale control. The input coordinates of sample points are transformed into Fourier domain frequency signals, which are filtered using high- and low-frequency thresholds corresponding to the maximum and minimum allowable feature sizes, l_{max} and l_{min}, respectively. For each sample point $p = (x, y)$, the frequency signals are computed as shown in Equation (3), and are then passed through the leaky ReLU activation function and the sigmoid normalization function before being fed into the neural network. The relationships between the frequency thresholds and the allowable feature sizes are shown in Equation (4).

$$\begin{aligned} FT_f(p) = \Big[& \cos\Big(\frac{\pi}{h}\big(f_1^x x + f_1^y v\big)\Big), \sin\Big(\frac{\pi}{h}\big(f_1^x x + f_1^y v\big)\Big) \\ & \cos\Big(\frac{\pi}{h}\big(f_2^x x + f_2^y v\big)\Big), \sin\Big(\frac{\pi}{h}\big(f_2^x x + f_2^y v\big)\Big) \\ & \vdots \\ & \cos\Big(\frac{\pi}{h}\big(f_{n_f}^x x + f_{n_f}^y v\big)\Big), \sin\Big(\frac{\pi}{h}\big(f_{n_f}^x x + f_{n_f}^y v\big)\Big)\Big] \end{aligned} \tag{3}$$

$$f \in \left[\frac{h}{l_{max}}, \frac{h}{l_{min}}\right] \tag{4}$$

where $f = \{f_1^x, f_1^y, f_2^x, f_2^y, f_{n_f}^x, f_{n_f}^y\}$ is the set of Fourier frequencies, n_f is the number of frequencies, $FT_f(p)$ is a $2n_f$-length vector after Fourier mapping, and h is the mesh edge length.

Structural compliance is evaluated by finite element analysis using a uniformly sampled grid. The forward computation of sample densities, static equilibrium equation, and computation of final compliance are given in Equations (5)–(7), respectively.

$$\rho_e(w, f, p)_{p \in e} = \rho_w(FT_f(p)) \tag{5}$$

$$J_e = \boldsymbol{u}_e^{\top} \boldsymbol{K}_e \boldsymbol{u}_e \tag{6}$$

$$J = \sum_e \rho_e^p J_e \tag{7}$$

where e is the analysis element, J_e is the compliance of element e, and J is the total compliance.

In the experiment of this study, an FEA is conducted with one sample point located at the center of element. When higher FEA accuracy is required, it is straightforward to use multiple sample points within an element, taking advantage of a continuous density function. The number and distribution of Fourier frequencies have a significant influence on design solutions; this will be analyzed and used to enhance solution diversity in Section 4.

3.3. Sketch-Based Constraints

Free-hand sketches are intuitive for designers to express ideas. Some aspects of sketch images such as line thickness or shading variations are irrelevant to the desired structural topology, and the sketch can be simply decomposed into open-stroke elements and closed-shape elements. Corresponding constraint terms tailored to the unique characteristics of strokes and closed shapes are introduced. Strokes reflect the designer's intended structural boundaries without imposing constraints on material distribution in adjacent regions. During the early design ideation, designers typically conceptualize characteristic lines that capture key stylistic or visual features rather than defining precise shapes. To accommodate this, a stroke-based constraint is introduced, ensuring that the structural boundaries of the generated design align with the sketched strokes while maintaining a degree of freedom in neighboring regions to facilitate further exploration of the design space. The proposed constraint is an undirected structural boundary constraint, where the loss is computed based on the similarity between current structural boundaries and the sketched strokes, as described in Equation (8).

$$L_{edge} = \sum_{i,j} \left(\boldsymbol{RI}_{i,j} - \overline{\boldsymbol{E}}_{i,j} \right)^2 \cdot \mathbb{I}_{[\boldsymbol{RI}_{i,j} > 0.5]} \tag{8}$$

where $\boldsymbol{RI}$ represents the sketch image normalized to [0,1], $\overline{\boldsymbol{E}}$ represents the current structural boundaries within the design domain, and $\mathbb{I}_{[\cdot]}$ is the indicator function.

The structural boundaries are extracted by a differentiable convolution operator on the density distribution map, where the convolution kernel is the edge-detecting Sobel operator, as shown in Equations (9)–(11). The density distribution map is obtained by forward computation of the density function at uniformly sampled grid points. In experiments, the sample grid resolution is taken as the image resolution of the sketch for convenience.

$$\boldsymbol{E} = \sqrt{(\boldsymbol{I} * \boldsymbol{K}_x)^2 + (\boldsymbol{I} * \boldsymbol{K}_y)^2 + \epsilon} \tag{9}$$

$$\boldsymbol{E}^* = \frac{\boldsymbol{E} - \min(\boldsymbol{E})}{\max(\boldsymbol{E}) - \min(\boldsymbol{E})} \tag{10}$$

$$\overline{\boldsymbol{E}} = \boldsymbol{E}^* \cdot \mathbb{I}_{[(i,j) \notin \boldsymbol{P}]} \tag{11}$$

where $\boldsymbol{I}$ is the density distribution map, $*$ is the convolution operator, $\boldsymbol{K}_x$ and $\boldsymbol{K}_y$ are the Sobel kernels along the x- and y-directions, respectively, ϵ is a small constant to avoid division by zero (in experiments, $\epsilon = 1 \times 10^{-6}$), and $\boldsymbol{P}$ is the non-design field within the density distribution map.

The sketch-based constraint is transformed into a penalty loss term for the neural density function. Since the steps of structural boundary extraction and constraint loss computation are differentiable, sensitivity analysis for this constraint term can be automated through the back-propagation of the neural network. In experiments, morphological dilation is applied to preprocess the sketch image to improve the saliency of strokes, and the weight coefficient of this constraint is gradually increased as similar to the volume constraint.

The closed-shape elements not only contain contour lines, which reflect the intended structural boundaries, but also signify the visual integrity of the drawn shape by the closure of the contours. Based on this, a direct way to apply a constraint is to make the closed shape directly represent a cavity or a solid. This study presents a more flexible constraint that drives the material distribution within the closed-shape region to be nearly uniform while allowing some variation. It is transformed to a penalty loss term, thereby enforcing a soft constraint to achieve visual integrity. The calculation of this loss term is shown in Equation (12).

$$L_{area} = \sum_{i,j} \left(\boldsymbol{RIF}_{i,j} - \boldsymbol{I}_{i,j} \right)^2 \cdot \mathbb{I}_{[\boldsymbol{RIF}_{i,j} > 0.5]} \tag{12}$$

where $\boldsymbol{RIF}$ is the preprocessed sketch image with the closed-shape region flood-filled and normalized to $[0, 1]$.

The weight coefficient for this loss term is a useful control parameter to adjust the saliency of the drawn closed shapes in the generated structure. When it is set as small, the generated structure may have small branches within the closed shape which are not visually obvious but significantly reduce the total compliance. Also, the preserved freedom of exploration within the closed shapes promotes solution diversity. When it is set as large, the closed-shape-based constraint approximates the hard boundaries of the design field, which is suitable when local structural details have been determined.

In application, designers can first conceptualize key stylistic structural features, sketch them, and make the algorithm explore diverse designs by setting a small weight coefficient for the closed-shape-based constraint. In subsequent interactive iterations between designers and the TO algorithm, structural parts are gradually designed and the constraint weight coefficient can be set to be larger for structural shapes in those parts to preserve them, until the structural design is complete.

The total loss function for updating the neural-encoded density function is given in Equation (13).

$$L = \frac{J}{J_0} + \alpha \left(\frac{\sum_e \rho_e v_e}{V*} - 1 \right)^2 + \beta \cdot L_{edge} + \gamma \cdot L_{area} \tag{13}$$

where J_0 is the initial total compliance value for normalization, α is the weight coefficient for the volume constraint, and β and γ are weight coefficients for the stroke-based constraint and closed-shape-based constraint, respectively.

3.4. Parallel Computing for Multi-Solution TO Framework

In the proposed TO process, perturbing the load force magnitudes, the initial weights of the neural network, as well as the Fourier sampling frequencies are all useful ways to generate multiple diverse designs. To account for the low interpretability of the neural network, this study perturbs the Fourier sampling frequencies $\boldsymbol{f}$ and the load force magnitudes $\boldsymbol{F}$ across multiple TO processes. As will be proved by the experiments in Section 4, this can provide sufficient design diversity.

A larger parallel scale increases the likelihood of discovering more innovative and high-performance designs, but it also increases the computing cost and produces more redundant solutions that are identical or highly similar to each other. To balance this conflict and remove redundant solutions, an adaptive parallel scale is employed. Multiple synchronized parallel TO processes are run, and a similarity measure for each pair of solutions within the current solution is set after a fixed number of iterations are conducted. A randomly selected solution is removed from pairs with high similarity to reduce the parallel scale before continuing the optimization iterations.

Similarity indexes include differences in compliance values and the image-based Structural Similarity Index Measure (SSIM) and pixel-wise mean absolute error (MAE), as computed by Equations (14) and (15). The SSIM index is better at capturing the structural characteristics that align with human visual perception, while the MAE index better reflects the data-level difference.

$$SSIM(x,y) = \frac{(2\mu_x\mu_y + C_1)\cdot(2\sigma_{xy} + C_2)}{\left(\mu_x^2 + \mu_y^2 + C_1\right)\cdot\left(\sigma_x^2 + \sigma_y^2 + C_2\right)}. \quad (14)$$

$$MAE(x,y) = \frac{1}{N}\sum_{i=1}^{N}|x_i - y_i| \quad (15)$$

where x and y represent the density distribution maps of two solutions, μ_x and μ_y are the average density values, σ_x^2 and σ_y^2 are density variances, σ_{xy} is the covariance of density distribution maps, C_1 and C_2 are small constants to avoid division by zero, and N is the number of pixels.

In early iterations of the TO process, the number of gray cells with intermediate density values is large, which makes the structural boundaries blurry. As the iterations progress and the SIMP penalty factor increases, the densities tend to binarize and structural boundaries become more clear and stable. As design diversity is mostly cultivated in the early iterations, a parallel scale is adapted only after the early iterations.

A conflict between human preferences expressed through sketches and the compliance minimization objective is very likely to occur. Guided by sketch-based constraints, some generated structural features may be detrimental to compliance minimization, thus are unreasonable for multi-objective optimization. Recognizing such structural features, i.e., the region of conflict, is important to ensure the quality of solutions.

To achieve this, we propose a sensitivity-based method to automatically recognize the region of conflict in structures. Given the TO framework's total loss minimization, the elements with positive sensitivity values for the sum of compliance loss and sketch-based penalty loss, denoted as L_t, should have their density values decreased under the virtual condition where sketch-based constraints are not applied. Therefore, among the solid elements in the generated structures, the ones with positive sensitivity values can be recognized as the region of conflict. The computation is expressed in Equation (16). The sensitivity can be conveniently computed via gradient back-propagation of neural networks, and the autograd function of PyTorch 1.5 is implemented in the experiment. During iterations of adaptive-scale parallel solving, solutions with large conflict regions are also removed.

$$F = \{\rho_e | \frac{\partial L_t}{\partial \rho_e}, \rho_e > 0.8\} \quad (16)$$

Algorithm 1 is the pseudo-code of the multi-solution TO framework with adaptive parallel scale.

Algorithm 1: Pseudo-code of multi-solution TO framework with human controllability and adaptive parallel scale.

Input: $V*$, **RI**, l_{min}, M
Parameter: ρ_w
Output: $\{TO_i\}$
1. Initialize $E_0, K_e^0, t_{SSIM}, t_{MAE}, t_{comp}, t_{vol}, N, N_0, n, \alpha, \beta, \gamma, p = 1$
2. Compute **RIF** from **RI**
3. For each of M individual TO process do:
4. Randomly initialize neural weights $\boldsymbol{w}$
5. Sample f according to Equation (1); $\boldsymbol{F}_b$, grid points $\boldsymbol{p}_{FEA}$ for FEA, $\boldsymbol{p}_{OP}$ for output
6. Compute J_0
7. For each iter of N iteration do:
8. For each of n parallel steps do:
9. For each i of M individual TO process do:
10. NN forward for $p_w(\boldsymbol{FT}_f(\boldsymbol{p}_{OP}))$, generate current design TO_i
11. NN forward for $p_w(\boldsymbol{FT}_f(\boldsymbol{p}_{FEA}))$, compute $J_i, L_i, V_i = \sum p_e v_e$
12. NN backward, update $\boldsymbol{w}$
13. Update $p = \min(p_{\max}, p + p_{\text{Inc}})$; $\alpha = \min(\alpha_{\max}, \alpha + \alpha_{\text{Inc}})$
14. $\beta = \min(\beta_{\max}, \beta + \beta_{\text{Inc}})$; $\gamma = \min(\gamma_{\max}, \gamma + \gamma_{\text{Inc}})$
15. If iter $> N_0$ then:
16. For each pair $(TO_i, TO_j)_{i \neq j}$ do:
17. Compute $SSIM(i,j)$, $MAE(i,j)$
18. If $SSIM(i,j) > t_{SSIM}$ and $MAE(i,j) < t_{MAE}$ and $|J_i - J_j| < t_{\text{comp}}$ and $|V_i - V_j| < t_{\text{vol}}$ then:
19. delete TO_j
20. For each TO_i do:
21. If TO_i has large region of conflict then:
22. delete TO_i
23. Return TO solutions $\{TO_i\}$

4. Experiments

4.1. Experiment Settings

This paper uses the innovative design of automotive wheels as a case study to validate the proposed multi-solution TO framework. The focus of the design is the disk-view 2D structural pattern of the spokes, which can be solved as a 2D structural TO problem. Typical design requirements for wheel hubs include being lightweight and having a high stiffness, using the structural yield (compliance) under specific static loading conditions as the stiffness metric. Accordingly, the TO problem is to minimize compliance under the total material volume constraint. Since the design of automotive wheels also requires aesthetic considerations, a balance between structural performance and aesthetics is desired, making this case highly suitable for validating the TO framework, aiming at enhancing solution diversity and human controllability.

Specific definitions of the design field and boundary conditions are shown in Figure 2. Structural compliance under both normal force and shear force is used as the stiffness metric. The normal force simulates the weight of the vehicle and wheel hub itself. It distributes on the outer rim within a 30° central angle range. The shear force simulates the ground friction force on the wheel, acting at the midpoint of the normal force distribution. Within a square domain, density values are set as 1 (solid) for the wheel rim and axle regions, variable (actual design field) for the spoke region, and 0 (void) for the other region. A fixation constraint is applied at the center of the axle.

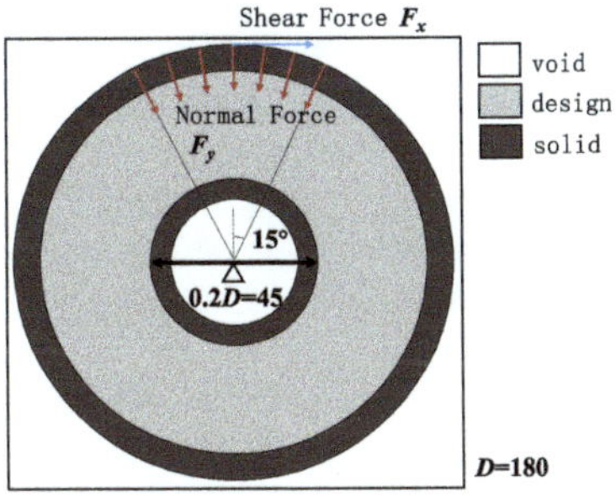

Figure 2. Definition of design field, boundary, and load conditions.

The rotational symmetry of the wheel spoke is handled explicitly: the annulus design field is evenly split into several fan units, with one (e.g., the topmost) designated as the reference unit. After sampling grid points for density computation, the points outside the reference unit are mapped to their central symmetric counterparts inside the reference unit, based on angular intervals defined by the fan angle. The counterparts are then taken as the actual input to the density function.

The content of this section illustrates the following: (1) an experiment for generating innovative structural designs by the TO framework; (2) an experiment for human controllability; (3) an analysis of the impacts of human control on both the design quality and diversity; (4) a comparative analysis with existing methods; and (5) an analysis of adaptive-scale parallel computing. The first experiment is used to verify the ability of the proposed method to enhance structural design exploration, and a qualitative evaluation of the generated results is conducted. The second experiment is used to verify that our method enables designers to control structural features precisely and flexibly. The third experiment is conducted to quantitatively analyze the impacts of human control parameters on design quality and diversity, in order to provide insights for designers to better fine-tune them. The fourth experiment is to verify the superiority of the proposed method over existing ones. The fifth experiment is to verify the usefulness of an adaptive scale in improving computing efficiency. Figure 3 gives a summary of the experiment design.

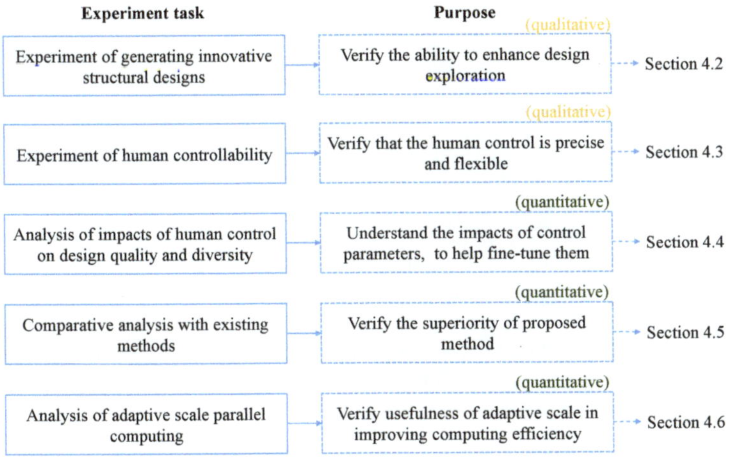

Figure 3. Summary of experiment design.

4.2. Experiment for Generating Innovative Structural Designs

The TO framework is tested using the hyper-parameters summarized in Table 1 to generate innovative structural designs of automobile wheels. α, β, and γ are initialized as 0. α_{Inc}, β_{Inc}, and γ_{Inc} are the increment values in each TO iteration. n_{FEA} is the grid sampling resolution for FEA computation. An adaptive parallel scale is not implemented by setting N_0 equal to N. Sketch images are preprocessed by applying a morphological dilation operator of kernel size (2,2). Figure 4 illustrates the loss curves and intermediate solutions during the iterative TO process. The two sketch-based constraint penalty losses take effect on the density distributions from the beginning of the iterations. As it iterates, the volume constraint has a stronger effect, with a decreased penalty loss value, the compliance gradually converges, while the sketch-based constraint penalty losses exhibit some degree of fluctuation. To ensure the comparability of results across multiple TO processes, a fixed number of iterations is set in the experiments. The convergence of each solving process is verified according to the relative changes in the loss terms for compliance and volume constraint. Over the last 30 iterations, if the relative changes in both losses remain below a threshold, the conversion of the process is validated.

Table 1. Hyper-parameters of multi-solution TO framework.

Parameter	Value	Parameter	Value	Parameter	Value
Desired volume	0.35	N_0	30	α_{max}	100
NN layers	$1 \times (100,)$	n	5	β_{max}	1
n_{EEA}	180	M	20	γ_{max}	0.1
l_{min}	9	α_{Inc}	0.3	p_{max}	6
l_{max}	80	β_{Inc}	5×10^{-4}	E	1
n_f	1000	γ_{Inc}	5×10^{-5}	Nu	0.3
n_f	30	p_{Inc}	0.04		

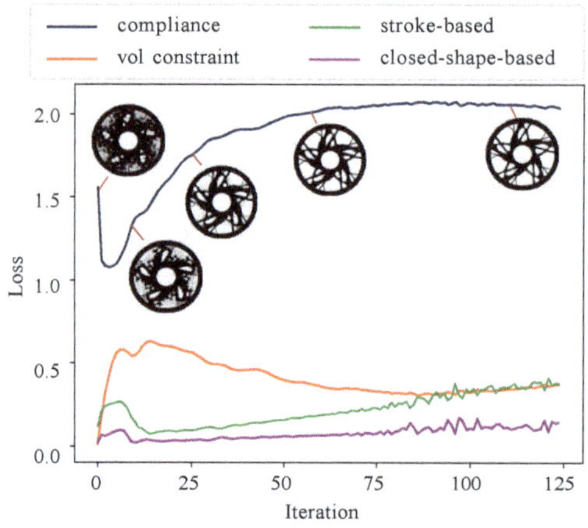

Figure 4. Loss curves and intermediate solutions during the iterative TO process.

Figures 5 and 6 show the generated innovative structural designs constrained by two examples of free-hand sketches. Using a parallel scale of $M = 20$ without an adaptive scale, a total of 20 designs are generated for each sketch image; from them, 8 randomly selected designs are displayed.

The sketch in Figure 5 contains sharp, angular strokes, reflecting a bold and austere style, while the sketch in Figure 6 displays droplet-shaped forms, reflecting a soft and graceful style. Qualitatively, it is evident that the generated solutions embody the visual styling features of those sketches. Also, they exhibit large differences from each other in terms of structural topology and geometric features, verifying the effectiveness of the proposed TO framework in generating innovative structural designs that conform to free-hand sketches and are diverse.

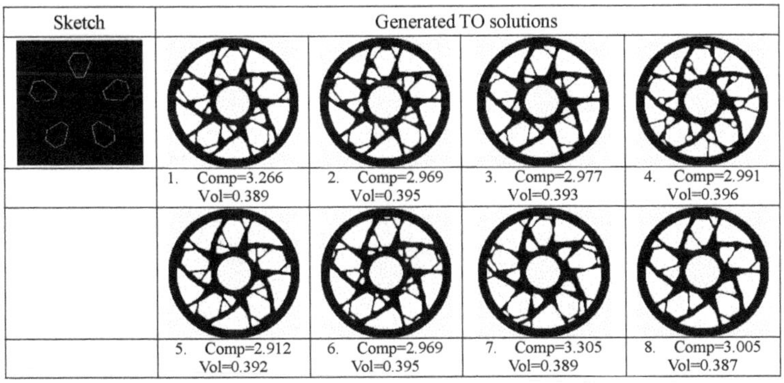

Figure 5. Generated innovative structural designs constrained by sketch image 1.

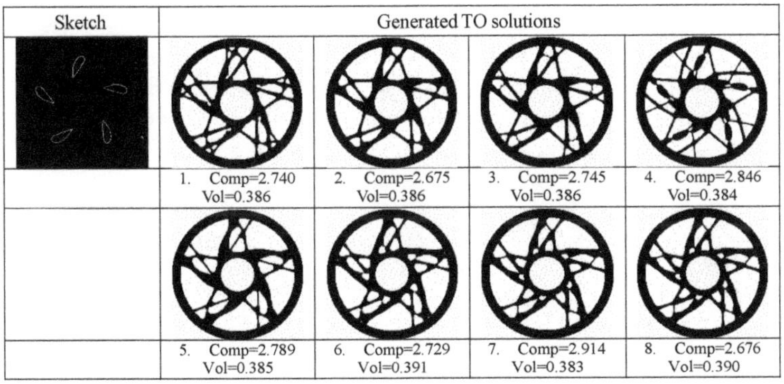

Figure 6. Generated innovative structural designs constrained by sketch image 2.

In most of the generated structures, a cavity that closely matches the closed-shape contour is generated, though subtle differences between them appear in some solutions. This is due to the closed-shape-based constraint being assigned with a large weight coefficient ($\gamma_{Inc} = 0.1\beta_{Inc}$). In a special case of solution 4 in Figure 6, the contour is matched by a solid structure rather than a cavity. This is caused by the undirected boundary constraint suppressing the closed-shape-based constraint. These experimental results show that the former constraint is beneficial for promoting solution diversity while achieving an effective sketch-based constraint. Also, the competition between two different constraint terms further facilitates the diverse design exploration.

Figure 7 presents more examples of generated solutions. Among them, two sketch images that only contain stroke elements also achieve effective human control. In solution 2 of Figure 7, the drawn strokes are implicitly reflected by several contour segments of different structural branches, demonstrating the flexibility of the undirected boundary constraints. Such flexibility benefits the generation of solutions with appealing visual effects. The compliance values vary by a large amount across the solution sets generated by different sketches, which indicates that the rationality of sketches has a decisive impact on the generated designs' quality. The volume ratios have very small variance, which is due to the increment in the volume constraint weight being much greater than that of the sketch-based constraints.

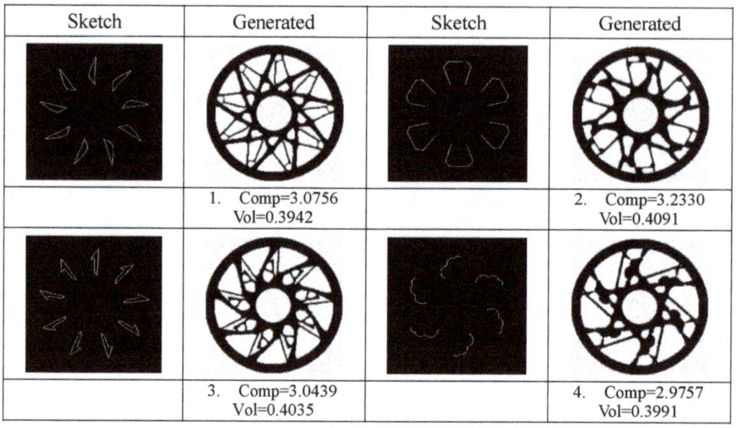

Figure 7. Generated innovative structural designs constrained by sketch images 3–6.

When random perturbation is applied to more hyper-parameters of the TO algorithm, more diverse and innovative structural designs can be generated. Here, experiments generating more broadly diversified innovative structural designs of wheel hubs are conducted. The desired volume, load force magnitude, and Fourier sampling frequencies are all randomly perturbed. Meanwhile, the generated TO design images are further converted into 3D models to facilitate intuitive evaluation by designers and subsequent detailed simulation analysis. The 2D-to-3D conversion steps are illustrated in Figure 8.

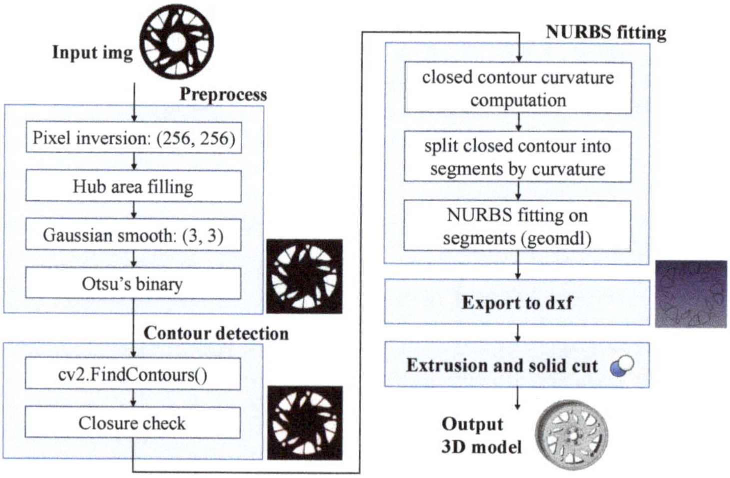

Figure 8. Illustration of 2D-to-3D conversion steps.

The input image is first preprocessed by pixel inversion, hub region filling, denoising with Gaussian smoothing, and binarization based on Otsu's method. Then, the contours within the design field are extracted and closed. After that, NURBS fitting is carried out on contour segments, which are split according to curvature. Finally, the NURBS curves are exported in the DXF format, and 3D models are obtained through extrusion and solid cutting operations.

Figure 9 showcases the innovative structural designs generated under the control of six different sketch examples, as well as the corresponding 3D design models. The obtained diverse innovative designs not only satisfy the sketch constraints but also have good performances in terms of the basic compliance metric. Their variance in shape is very useful to stimulate the creative thinking of designers. In practice, designers can first evaluate the generated solutions qualitatively in terms of aesthetic appeal, innovativeness, etc., then choose the ones with low compliance and human preference to further carry out a multi-criteria mechanical simulation analysis.

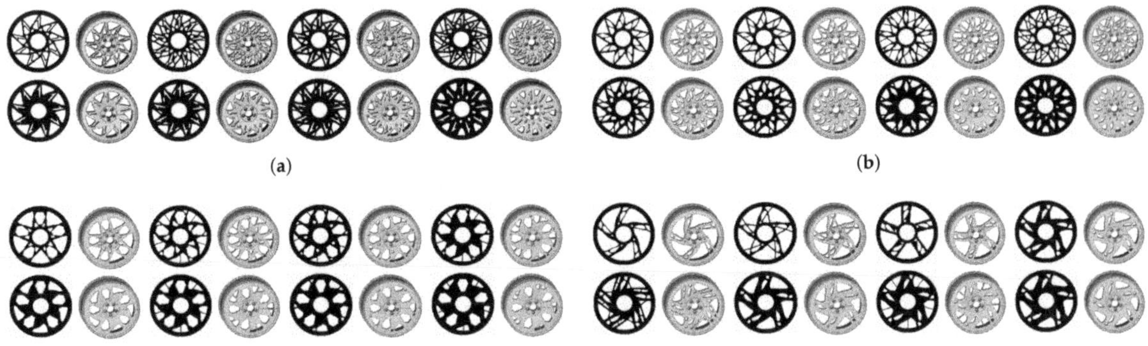

(**a**) (**b**) (**c**) (**d**)

Figure 9. Innovative design exploration results for automobile wheel spokes under sketch-based constraints. Randomly perturbed hyper-parameters include dv, $l_{\min}$, and **f**. (**a**) Guided by sketch 1. (**b**) Guided by sketch 2. (**c**) Guided by sketch 3. (**d**) Guided by sketch 4.

4.3. Experiment of Human Controllability

Figure 10 presents the experimental results of human controllability over the generated structural features by integrating the minimum allowed structural feature size l_{min} and the relative weight coefficient of the closed-shape-based constraint over the stroke-based constraint λ ($\gamma_{Inc} = \gamma \cdot \beta_{Inc}$). It is intuitive for designers to take control by simply fine-tuning these two hyper-parameters, as their control effects are precisely targeting at specific structural features and are independent from each other.

When the control parameter λ is set to be small, the drawn closed-shape region contains many branching structures. In this case, if the value of l_{min} is also small, the generated structure is highly complex, with low manufacturability, the drawn shape is not visually prominent, though the compliance is small; if l_{min} is larger, the structural complexity becomes lower, the hand-drawn shape is more prominent, but compliance increases significantly. When λ is set to be larger, there are fewer branching structures within the drawn shape region, leading to a result similar to a hard constraint on the design field boundary. The shape becomes more prominent, especially when the shape has a large area, which significantly affects the aesthetic styling. l_{min} has a bigger impact on compliance than λ. These experimental results verify that the two control parameters have different effects on the structural compliance, and mechanical and aesthetic performance, enabling human designers to achieve a better balance between multiple factors in a flexible control manner.

The flexibility of human controllability is valuable for practical application, as solving conflicts between different design factors is the main challenge in conceptual structural design. Mechanical and aesthetic performances are conflicting in most cases. Meanwhile, increased structural complexity lowers the manufacturability, as overly narrow structural branches may be unfeasible to produce. Structural complexity also impacts aesthetic appeal, with simpler and regular forms generally considered more aligned with modern aesthetic standards.

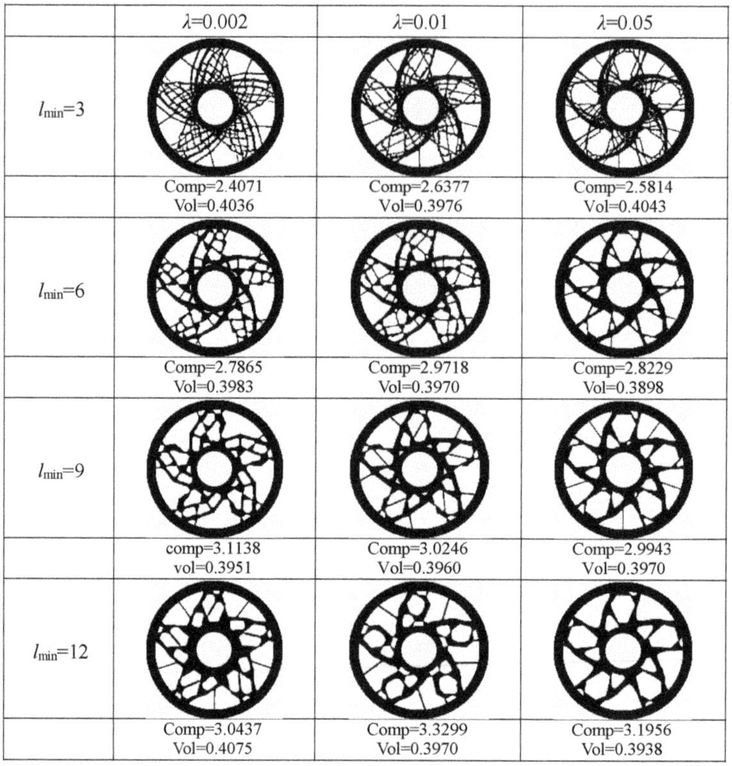

Figure 10. Experimental results of human controllability by integrating l_{min} and λ.

4.4. Impacts of Human Control on Design Quality and Diversity

The qualitative analysis of the experimental results as above has verified the effectiveness of the proposed TO framework to generate conceptual structural designs while providing flexible human controllability. The results also indicate that human control may conflict with the compliance minimization objective of the TO algorithm, leading to an increase in the compliance minimization. In the following, an ablation study is presented that quantitatively analyzes the impact of the three human control approaches on both the solution quality and diversity.

The design quality metrics include the total structural compliance and volume. Although multiple aspects in mechanics, fatigue life, heat transfer, manufacturability, etc., are necessary to verify the engineering applicability of design, it is too costly and time-consuming to carry out a detailed analysis for all design in the conceptual stage. Thus, using compliance and volume criteria can effectively find the initial designs that meet the basic requirements.

Diversity metrics include the image-based average SSIM index and MAE index between pairs of solutions, denoted as $\overline{SSIM}$ and $\overline{MAE}$, respectively, and the coefficients of variance for the compliance and volume ratio, denoted as CV_{comp} and CV_{vol}. Their computation formulas are given in Equations (16)–(18).

$$\overline{SSIM} = \frac{\sum_{i \neq j} SSIM(i,j)}{M} \tag{17}$$

$$\overline{MAE} = \frac{\sum_{i \neq j} MAE(i,j)}{M} \tag{18}$$

$$CV_* = \frac{\sigma(*)}{\mu(*)} \times 100\% \tag{19}$$

where $\sigma(*)$ is the standard deviation of the metrics' values, and $\mu(*)$ is the mean of the metrics' values.

In the ablation experiment of the stroke-based constraint, parameter β_{Inc} is selected from $[0, 5 \times 10^{-5}, 2 \times 10^{-4}, 5 \times 10^{-4}, 8 \times 10^{-4}]$; a closed-shape-based constraint is not applied ($\lambda = 0$). In the ablation experiment of the closed-shape-based constraint, as contour lines are essential to make the drawn shape visually prominent, β_{Inc} is set to 5×10^{-4}, and λ is selected from $[8 \times 10^{-5}, 4 \times 10^{-4}, 2 \times 10^{-3}, 1 \times 10^{-2}, 5 \times 10^{-2}]$. In the ablation experiment of the minimal allowed feature size, l_{min} is selected from [3, 6, 9, 12], β_{Inc} is set to 5×10^{-4}, and λ is set 0.05. The other hyper-parameters are the same as in Table 1. The sketch images of Figures 5 and 6 are utilized as case 1 and case 2.

The experimental results of the ablation study are shown in Figures 11–14. Quantitative results of several TO algorithm runs are given in Table 2. An increase in β_{Inc} leads to obviously larger compliance. The average compliance value is 2.62 when this constraint is not applied, but increases to 3.15 and 2.87 for case 1 and case 2, respectively, given $\beta_{Inc} = 5 \times 10^{-4}$. Importantly, an increase in β_{Inc} does not reduce the solution diversity but slightly improves it. A possible reason is that this constraint makes it easier to escape local optima through interfering with the compliance minimization objective while avoiding limiting the material distribution in the regions surrounding the strokes, thereby exploring the design space more divergently. Therefore, the undirected structural boundary constraint is very beneficial to preserve freedom of design exploration under sketch-based human control.

In contrast, the experimental results reveal that the closed-shape-based constraint reduces the solution diversity but has no obvious impact on the compliance or volume ratio. From an analysis of the results in Figure 10, the main reason for the reduction in the solution diversity is that a large λ makes the drawn shape region approximate a non-design field. Stable compliance and volume ratio values indicate that the "remaining" design field is still sufficient for exploration. It is verified that by decomposing sketches into strokes and closed-shape-based elements with distinct control features, the solution quality and diversity can also be better balanced.

The experimental results also reveal that an increase in the minimal allowed feature size leads to larger compliance and reduced diversity, which aligns with the conclusions drawn from the qualitative analysis. Based on the control features of the three control parameters, a proper approach is to first focus on drawing strokes and simply rely on a stroke-based constraint to promote diverse exploration; subsequently, add closed shapes and increase this constraint weight incrementally to optimize compliance while restricting the material volume. The minimal allowed feature size should be preferably set to a large value according to multifactor requirements in practice.

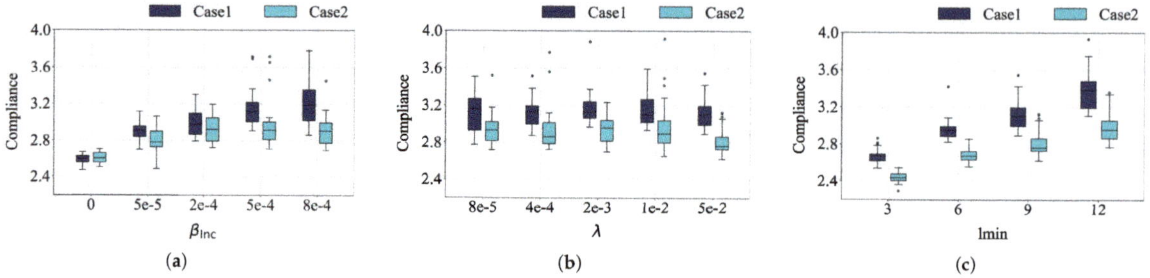

Figure 11. Ablation study of β_{Inc} (**a**), λ (**b**), and l_{min} (**c**) in terms of compliance.

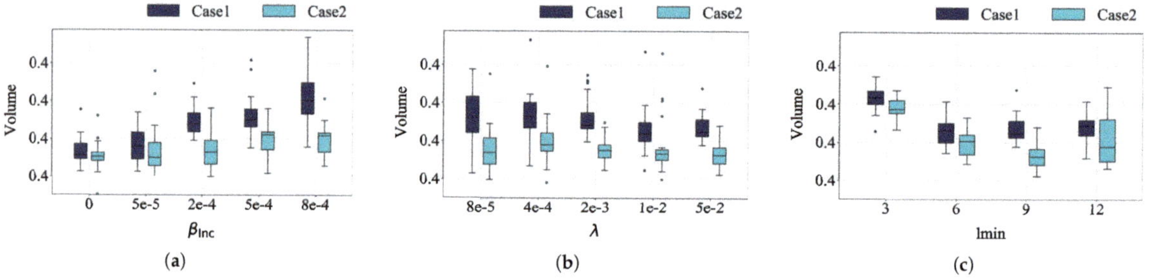

Figure 12. Ablation study of β_{Inc} (**a**), λ (**b**), and l_{min} (**c**) in terms of volume.

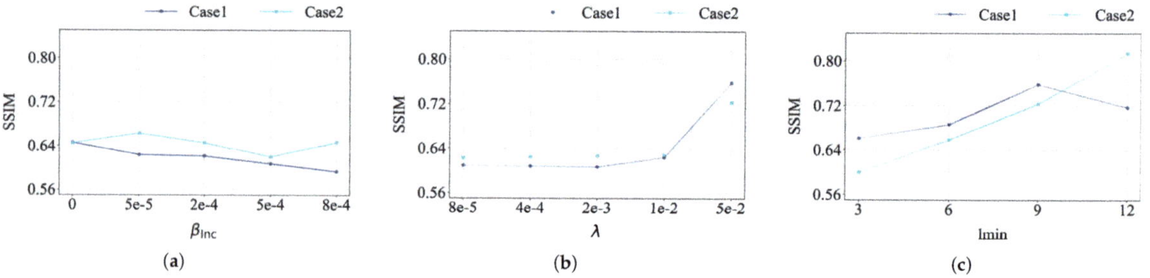

Figure 13. Ablation study of β_{Inc} (**a**), λ (**b**), and l_{min} (**c**) in terms of $\overline{SSIM}$.

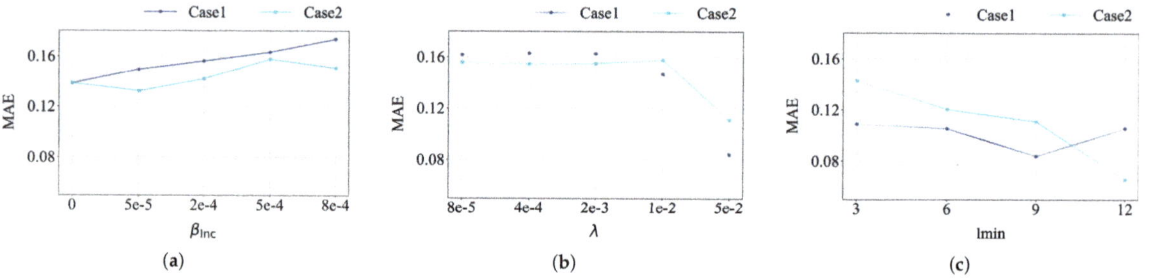

Figure 14. Ablation study of β_{Inc} (**a**), λ (**b**), and l_{min} (**c**) in terms of $\overline{MAE}$.

Table 2. Quantitative experimental results of ablation study.

Case 1						
Setting	μ_{comp}	μ_{vol}	CV_{comp}	CV_{vol}	$\overline{SSIM}$	$\overline{MAE}$
$\beta_{Inc} = 0$	2.5831	0.3892	0.0204	0.0074	0.6461	0.1385
$\beta_{Inc} = 5 \times 10^{-4}$	3.1602	0.3970	0.0687	0.0111	0.6069	0.1634
$\beta_{Inc} = 5 \times 10^{-4}, \lambda = 0.05$	3.1263	0.3948	0.0523	0.0073	0.7585	0.0841
$\beta_{Inc} = 5 \times 10^{-4}, \lambda = 0.05, l_{min} = 12$	3.4158	0.3952	0.0798	0.0068	0.7170	0.1058
Case 2						
Setting	μ_{comp}	μ_{vol}	CV_{comp}	CV_{vol}	$\overline{SSIM}$	$\overline{MAE}$
$\beta_{Inc} = 0$	2.5979	0.3884	0.0230	0.0079	0.6457	0.1386
$\beta_{Inc} = 5 \times 10^{-4}$	2.9896	0.3919	0.0939	0.0094	0.6199	0.1579
$\beta_{Inc} = 5 \times 10^{-4}, \lambda = 0.1$	2.8207	0.3894	0.0558	0.0070	0.7235	0.1111
$\beta_{Inc} = 5 \times 10^{-4}, \lambda = 0.1, l_{min} = 12$	2.9787	0.3928	0.0565	0.0138	0.8138	0.0659

4.5. Comparative Analysis with Existing Methods

The proposed TO framework is compared with the sketch-guided TO algorithm (baseline 1) by Oh et al. [26] and the interactive TO algorithm (baseline 2) by Li et al. [12]. Baseline 1 directly utilizes the entire reference sketch image and a similarity constraint between the TO solution and the sketch. In order to avoid using existing design datasets, to ensure comparability between methods, load condition parameter perturbation is adopted for baseline 1, replacing its data-driven approach of enhancing diversity. Baseline 2 combines the free-hand sketch and human scoring to achieve controllability, and achieves solution diversity by perturbing the sensitivity numbers of the soft-skilled BESO algorithm according to human input. To avoid the uncertainty of human input, the random perturbation of density values is employed as an approximate strategy, with three different perturbation magnitudes for analysis. The proposed TO framework without load condition parameter perturbation is also compared.

While traditional TO algorithms, such as SIMP or BESO, are unable to offer design diversity, existing multi-solution TO algorithms achieve this with three typical methods: simply perturbing the design parameters like load force magnitudes, data-driven feature learning, and adjusting the optimizable variables based on interactive human input. Data-driven methods offer strong design diversity, but both the diversity and quality of the solutions depend on the size and quality of the training dataset. As such, quantitative comparison between these two types of methods is unfeasible. Due to the uncertain effect of the subjectivity of human input, quantitative comparison between these types of methods and parameter perturbation-based ones is also unfeasible. Therefore, this paper only conducts quantitative comparative experiments using the two latest parameter perturbation-based methods, with the same optimizable parameters.

For each compared method, sketch example case 1 is used as the human input, and the perturbation range of the load force parameter F_x is set as [0.3, 1.5]. The evaluation criteria for the methods are the same as above, with the average of 20 repeated runs for each method being the evaluation metric.

Table 3 presents the results of the comparative experiment. In terms of solution quality, the proposed TO framework, along with baseline 1, outperforms baseline 2 by a large margin, verifying that perturbations in the load condition or Fourier sampling frequencies have a very similar impact on the solution quality, but perturbing the density values significantly reduces the quality. The proposed method achieves the best performances, as indicated by the metrics in bold. In terms of diversity, the proposed framework outperforms baseline 1, as well as outperforming baseline 2 when it has a small perturbation amplitude (<0.4). It only slightly lags behind baseline 2 when the perturbation amplitude is large (=0.4) in terms of morphological diversity. For baseline 2, as the amplitude increases, the compliance and volume increase significantly, even violating the volume constraint, as indicated by the μ_{comp} metric shown in bold. While baseline 2 can generate the most diverse solutions ($\overline{SSIM} = 0.7358$, $\overline{MAE} = 0.0663$), it excessively sacrifices solution quality and struggles to reasonably balance the conflict between solution quality and diversity by only adjusting the perturbation amplitude. The above comparative analysis verifies the contribution of the proposed framework in achieving both high design quality and diversity.

Table 3. Comparative experiment results for proposed TO framework and alternatives.

Setting	μ_{comp}	μ_{vol}	CV_{comp}	CV_{vol}	$\overline{SSIM}$	$\overline{MAE}$
Density perturb (amp = 0)	**3.1164**	0.3960	0.0151	0.0033	0.9731	0.0097
Density perturb (amp = 0.1)	**3.3117**	0.4019	0.0204	0.0047	0.9084	0.0207
Density perturb (amp = 0.4)	**3.8991**	0.4382	0.0192	0.0049	0.7358	0.0663
Load cond perturb	3.0322	0.3955	0.0243	0.0067	0.8700	0.0399
Fourier freq perturb	3.0322	0.3955	0.0298	0.0067	0.8641	0.0411
Proposed	**3.0322**	**0.3955**	**0.0317**	**0.0075**	**0.8339**	**0.0516**

4.6. Analysis of Adaptive-Scale Parallel Computing

To test the effectiveness of the adaptive parallel scale strategy for the multi-solution TO framework, comparative experiments are conducted with and without this strategy. Hyper-parameters $N_0 = 10$ and $N = 30$ when they are implemented, and $N_0 = N = 30$ when they are not implemented, with solution set filtering only after the iteration ends. The initial parallel scale M is set to 50 with the adaptive parallel scale, and selected from [50, 40, 30] without it. The sketch example of case 2 is utilized as the human input for all cases. The threshold values for the similarity metrics are fixed during the iterations, which ensures that solution pairs are retained as long as their difference is large according to the absolute criteria. Those values are first empirically set to ensure that about 50% of the solutions from the initial solution set are retained, then adjusted to make each similarity metric effective for detecting similar pairs. We set $t_{SSIM} = 0.9$, $t_{MAE} = 0.02$, $t_{comp} = 0.1$, and $t_{vol} = 0.01$.

A comparison of the experimental results is shown in Table 4. The number of retained solutions after filtering with the adaptive parallel scale is 16, the same as the number of obtained solutions without this strategy and a smaller initial parallel scale ($M = 30$). We highlight the results obtained using the two experimental settings in bold in Table 4, as their comparison validates the superiority of adaptive-scale parallel computing. However, this strategy leads to a shorter computing time and improved solution diversity, due to divergent exploration primarily occurring in the early optimization iterations, making a larger initial parallel scale more beneficial.

As described in Section 3, due to the conflict between sketch-based constraints and the compliance minimization objective, the generated structures may include some unreasonable features. Figure 15 showcases two examples that include unreasonable features generated under the guidance of the sixth sketch example. In Figure 15a, a very thin protruding branch is generated guided by the stroke-based constraint, but it is not connected to the other side of the structure. From the von Mises stress map, which is useful to visualize the distribution of equivalent stress within the structural field and is computed by the finite element analysis steps as in the TO algorithm, it is clear that this branch is not contributing as a load path, which suggests that it is an unreasonable design. In Figure 15b, the protrusion part on the support structure also does not contribute as load-bearing and exists solely to conform to the drawn shape of the sketch.

Table 4. Comparative experimental results for adaptive parallel scale method.

Setting	Time (s)	Num	μ_{comp}	μ_{vol}	CV_{comp}	CV_{vol}	$\overline{SSIM}$	$\overline{MAE}$
N/Ad M = 50	3060.7	50	2.7101	0.3887	0.0264	0.0058	0.8428	0.0488
filtered	/	23	2.7320	0.3895	0.0320	0.0063	0.8110	0.0601
N/A M = 40	2448.6	40	2.7164	0.3888	0.0253	0.0060	0.8453	0.0477
filtered	/	19	2.7391	0.3897	0.0313	0.0065	0.8170	0.0578
N/Ad M = 30	**1836.4**	30	2.7190	0.3888	0.0273	0.0058	0.8473	0.0472
filtered	/	**16**	**2.7415**	**0.3894**	**0.0330**	**0.0065**	**0.8241**	**0.0555**
Ad N = 50	**1810.4**	**16**	**2.7267**	**0.3891**	**0.0278**	**0.0054**	**0.8041**	**0.0617**
filtered	/	16	/	/	/	/	/	/

(**a**) unreasonable example 1 (**b**) unreasonable example 2

Figure 15. Unreasonable generated structures due to conflicts between sketch-based constraints, volume constraint, and compliance.

Figure 16 shows examples of recognized regions of conflict using the sensitivity-based method. The regions of conflict are marked in red. The thin protruding branch in the second solution is recognized as unreasonable. It is clear that this feature would be the first to be removed if only compliance minimization and the volume constraint were applied. This solution is removed from the solution set, while the first solution with a smaller region of conflict is retained.

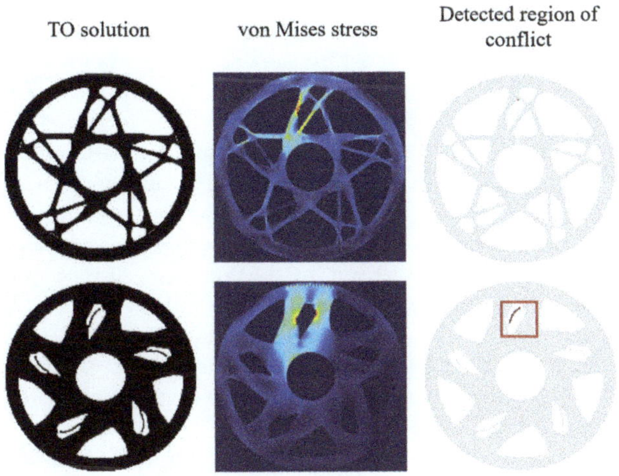

Figure 16. Examples of recognized regions of conflict.

5. Conclusions

This paper proposes a novel sketch-constrained multi-solution topology optimization algorithm, designed for human–machine collaborative structural conceptual design exploration scenarios. It allows designers to express preliminary structural design features through hand-drawn sketches, which form constraints for the topology optimization algorithm. The algorithm divides the sketch content into open-line elements and closed-shape elements, and introduces line-based undirected structural boundary constraints and closed-shape-based regional material distribution constraints. The former imposes no material distribution restrictions on the areas on either side of the lines, enabling the exploration of many novel forms that integrate the sketch's line elements into the topology, creating unique visual effects. Additionally, compared to unconstrained topology optimization, it does not reduce the morphological diversity of the design solutions. The latter drives uniform material distribution within the shape regions, which, while somewhat reducing the design exploration space and diversity, enhances the visual prominence of the sketch's shape elements. Importantly, when line constraints are already in place, this approach does not reduce structural performance, allowing designers to gradually refine local structural designs while exploring other parts of the structure in a high-quality, divergent manner. Furthermore, the mutual suppression between the two sketch constraints can further enhance design diversity.

In contrast, existing sketch constraint methods only directly use the overall sketch similarity constraint to control the topology optimization results. The experiments show that the proposed method provides designers with a finer and more flexible approach to controlling algorithm results through hand-drawn sketches, allowing for better balancing of structural performance, lightweight, visual effects, design exploration diversity, and novelty. Additionally, the design diversity offered by the proposed method is significantly stronger.

This algorithm also integrates Fourier mapping-based feature size control, allowing designers to combine it with sketch constraints for flexible control over both the local features and overall complexity of the generated structure. By introducing randomness through perturbations in the Fourier sampling frequency and adopting a parallel optimization strategy based on adaptive scale adjustment, the algorithm generates diversified and innovative design solutions. The experiments show that the proposed Fourier sampling frequency perturbation method outperforms existing boundary condition parameter perturbation and design domain density perturbation methods both in terms of design solution quality and diversity. Based on the proposed parallel optimization strategy, the multi-solution topology optimization algorithm in this paper offers comprehensive advantages in manual controllability, design solution quality, diversity, and computational efficiency, providing an effective solution for current topology optimization algorithms that are unsuitable for structural innovation design exploration.

The research and experimental process also identified several issues that need further resolution. First, the conflict between the sketch constraints reflecting human preferences and the algorithm's structural performance optimization objective has not been fully mitigated. Although the proposed algorithm uses sensitivity-based method to detect and remove unreasonable designs from the solution set, it is better to automatically generate solutions in which the material in each part of the structure not only satisfies the sketch constraints but also contributes to mechanical performance. Also, finite element calculations are the most time-consuming step in topology optimization algorithms, and exploring ways to avoid or accelerate the finite element computation process is also a valuable area for future research. Further, how to help designers quickly learn the basics of the control characteristics of hyper-parameters should be studied when the design structure becomes more complex and more hyper-parameters are included according to constraints like manufacturability. To improve the applicability of the proposed method in engineering design practice, future study will focus on involving the manufacturing constraints and consider the complex effects of material performance variance in different working conditions.

Author Contributions: Conceptualization, S.Z., J.H. and J.Q.; methodology, investigation, data, and writing, S.Z.; supervision, project administration, J.H. and J.Q.; funding acquisition, L.W., J.G., J.M. and G.Z. All authors have read and agreed to the published version of the manuscript.

Funding: This research is supported by National Natural Science Foundation of China (52475270, 52035007, U23B20102), and Xie Youbai Design Scientific Research Foundation (XYB-DS-202401).

Institutional Review Board Statement: Not applicable.

Informed Consent Statement: Not applicable.

Data Availability Statement: Code, data and the user interface for creating sketch image is publicly available at https://github.com/Sheeryer/SketchTO.git (accessed on 28 February 2025).

Conflicts of Interest: The authors declare no conflicts of interest.

References

1. Amir, O.; Sigmund, O. Reinforcement layout design for concrete structures based on continuum damage and truss topology optimization. *Struct. Multidiscip. Optim.* **2013**, *47*, 157–174. [CrossRef]
2. Sigmund, O.; Maute, K. Topology optimization approaches: A comparative review. *Struct. Multidiscip. Optim.* **2013**, *48*, 1031–1055. [CrossRef]
3. Oh, S.; Jung, Y.; Lee, I.; Kang, N. Design automation by integrating generative adversarial networks and topology optimization. In Proceedings of the ASME International Design Engineering Technical Conferences and Computers and Information in Engineering Conference, Quebec City, QC, Canada, 26–29 August 2018; Volume 2A, p. V02AT03A008.
4. Wu, J.; Clausen, A.; Sigmund, O. Minimum compliance topology optimization of shell-infill composites for additive manufacturing. *Comput. Methods Appl. Mech. Eng.* **2017**, *326*, 358–375. [CrossRef]
5. Zhu, J.-H.; Zhang, W.-H.; Xia, L. Topology Optimization in Aircraft and Aerospace Structures Design. *Arch. Comput. Methods Eng.* **2016**, *23*, 595–622. [CrossRef]
6. Alexandersen, J.; Sigmund, O.; Aage, N. Large scale three-dimensional topology optimisation of heat sinks cooled by natural convection. *Int. J. Heat Mass Transf.* **2016**, *100*, 876–891. [CrossRef]
7. Dilgen, S.B.; Dilgen, C.B.; Fuhrman, D.R.; Sigmund, O.; Lazarov, B.S. Density based topology optimization of turbulent flow heat transfer systems. *Struct. Multidiscip. Optim.* **2018**, *57*, 1905–1918. [CrossRef]
8. Dilgen, C.B.; Dilgen, S.B.; Fuhrman, D.R.; Sigmund, O.; Lazarov, B.S. Topology optimization of turbulent flows. *Comput. Methods Appl. Mech. Eng.* **2018**, *331*, 363–393. [CrossRef]
9. Albrechtsen, M.; Lahijani, B.V.; Christiansen, R.E.; Nguyen, V.T.H.; Casses, L.N.; Hansen, S.E.; Stenger, N.; Sigmund, O.; Jansen, H.; et al. Nanometer-scale photon confinement in topology-optimized dielectric cavities. *Nat. Commun.* **2022**, *13*, 6281. [CrossRef]
10. Christiansen, R.E.; Sigmund, O. Inverse design in photonics by topology optimization: Tutorial. *J. Opt. Soc. Am. B-Opt. Phys.* **2021**, *38*, 496–509. [CrossRef]
11. Sotiropoulos, S.; Kazakis, G.; Lagaros, N.D. Conceptual design of structural systems based on topology optimization and prefabricated components. *Comput. Struct.* **2020**, *226*, 106136. [CrossRef]
12. Li, Z.; Lee, T.-U.; Xie, Y.M. Interactive Structural Topology Optimization with Subjective Scoring and Drawing Systems. *Comput. Aided Des.* **2023**, *160*, 103532. [CrossRef]
13. Jang, S.; Yoo, S.; Kang, N. Generative Design by Reinforcement Learning: Enhancing the Diversity of Topology Optimization Designs. *Comput.-Aided Des.* **2022**, *146*, 103225. [CrossRef]
14. Chandrasekhar, A.; Suresh, K. Approximate Length Scale Filter in Topology Optimization using Fourier Enhanced Neural Networks. *Comput.-Aided Des.* **2022**, *150*, 103277. [CrossRef]
15. Ha, D.Q.; Carstensen, J.V. Human-Informed Topology Optimization: Interactive application of feature size controls. *Struct. Multidiscip. Optim.* **2023**, *66*, 59. [CrossRef]
16. Schiffer, G.; Ha, D.Q.; Carstensen, J.V. HiTop 2.0: Combining topology optimisation with multiple feature size controls and human preferences. *Virtual Phys. Prototyp.* **2023**, *18*, e2268603. [CrossRef]
17. Zhou, M.; Lazarov, B.S.; Wang, F.; Sigmund, O. Minimum length scale in topology optimization by geometric constraints. *Comput. Methods Appl. Mech. Eng.* **2015**, *293*, 266–282. [CrossRef]

18. Pellens, J.; Lombaert, G.; Lazarov, B.; Schevenels, M. Combined length scale and overhang angle control in minimum compliance topology optimization for additive manufacturing. *Struct. Multidiscip. Optim.* **2019**, *59*, 2005–2022. [CrossRef]
19. Zhang, Z.; Jiang, L.; Yarlagadda, T.; Zheng, Y.; Usmani, A. A novel multi-pattern control for topology optimization to balance form and performance needs. *Eng. Struct.* **2024**, *303*, 117581. [CrossRef]
20. Meng, X.; Zhang, L.-Y.; Zhao, Z.-L.; Xie, Y.M. A direct approach to achieving efficient free-form shells with embedded geometrical patterns. *Thin-Walled Struct.* **2023**, *185*, 110559. [CrossRef]
21. Li, Y.; Zhang, Z.; Luo, J.; Peng, W.; Zhou, W.; Yao, W. Concurrent topology optimization of shells with pattern-guided infills for intuitive design and additive manufacturing. *Comput. Methods Appl. Mech. Eng.* **2024**, *418*, 116485. [CrossRef]
22. Zhang, J.; Zhang, H.; Chen, J.; Liu, T.; Peng, J.; Zhang, D.; Yin, S. Topology optimization of periodic mechanical structures with orthotropic materials based on the element-free Galerkin method. *Eng. Anal. Bound. Elem.* **2022**, *143*, 383–396. [CrossRef]
23. Zhang, W.; Wang, Y.; Du, Z.; Liu, C.; Youn, S.-K.; Guo, X. Machine-learning assisted topology optimization for architectural design with artistic flavor. *Comput. Methods Appl. Mech. Eng.* **2023**, *413*, 116041. [CrossRef]
24. Vulimiri, P.S.; Deng, H.; Dugast, F.; Zhang, X.; To, A.C. Integrating Geometric Data into Topology Optimization via Neural Style Transfer. *Materials* **2021**, *14*, 4551. [CrossRef] [PubMed]
25. Zhang, W.; Wang, Y.; Youn, S.-K.; Guo, X. Machine learning powered sketch aided design via topology optimization. *Comput. Methods Appl. Mech. Eng.* **2024**, *419*, 116651. [CrossRef]
26. Oh, S.; Jung, Y.; Kim, S.; Lee, I.; Kang, N. Deep Generative Design: Integration of Topology Optimization and Generative Models. *J. Mech. Des.* **2019**, *141*, 111405. [CrossRef]
27. Yoo, S.; Lee, S.; Kim, S.; Hwang, K.H.; Park, J.H.; Kang, N. Integrating deep learning into CAD/CAE system: Generative design and evaluation of 3D conceptual wheel. *Struct. Multidiscip. Optim.* **2021**, *64*, 2725–2747. [CrossRef]
28. Xia, Z.; Wang, J.; Li, Y.; Zhang, L.; Liu, C. Intelligent design of key joints in aerial building machine using topology optimization and generative adversarial network. *Autom. Constr.* **2024**, *168*, 105747. [CrossRef]
29. Wang, M.Y.; Wang, X.M.; Guo, D.M. A level set method for structural topology optimization. *Comput. Methods Appl. Mech. Eng.* **2003**, *192*, 227–246. [CrossRef]
30. Bendsoe, M.P.; Sigmund, O. Material interpolation schemes in topology optimization. *Arch. Appl. Mech.* **1999**, *69*, 635–654. [CrossRef]
31. Bruns, T.E.; Tortorelli, D.A. Topology optimization of non-linear elastic structures and compliant mechanisms. *Comput. Methods Appl. Mech. Eng.* **2001**, *190*, 3443–3459. [CrossRef]
32. Zegard, T.; Paulino, G.H. Bridging topology optimization and additive manufacturing. *Struct. Multidiscip. Optim.* **2016**, *53*, 175–192. [CrossRef]
33. Vantyghem, G.; De Corte, W.; Shakour, E.; Amir, O. 3D printing of a post-tensioned concrete girder designed by topology optimization. *Autom. Constr.* **2020**, *112*, 103084. [CrossRef]
34. Li, Q.; Liang, G.; Luo, Y.; Zhang, F.; Liu, S. An explicit formulation for minimum length scale control in density-based topology optimization. *Comput. Methods Appl. Mech. Eng.* **2023**, *404*, 115761. [CrossRef]
35. He, Y.; Zhao, Z.-L.; Lin, X.; Xie, Y.M. A hole-filling based approach to controlling structural complexity in topology optimization. *Comput. Methods Appl. Mech. Eng.* **2023**, *416*, 116391. [CrossRef]
36. Lazarov, B.S.; Wang, F.; Sigmund, O. Length scale and manufacturability in density-based topology optimization. *Arch. Appl. Mech.* **2016**, *86*, 189–218. [CrossRef]
37. Zhong, S.; Punpongsanon, P.; Iwai, D.; Sato, K. Topology optimization with text-guided stylization. *Struct. Multidiscip. Optim.* **2023**, *66*, 256. [CrossRef]
38. Radford, A.; Kim, J.W.; Hallacy, C.; Ramesh, A.; Goh, G.; Agarwal, S.; Sastry, G.; Askell, A.; Mishkin, P.; Clark, J. Learning transferable visual models from natural language supervision. In Proceedings of the 38th International Conference on Machine Learning, Virtual, 18–24 July 2021; pp. 8748–8763.
39. Pouyanfar, S.; Sadiq, S.; Yan, Y.; Tian, H.; Tao, Y.; Reyes, M.P.; Shyu, M.-L.; Chen, S.-C.; Iyengar, S.S. A Survey on Deep Learning: Algorithms, Techniques, and Applications. *ACM Comput. Surv.* **2019**, *51*, 92. [CrossRef]
40. Nie, Z.; Lin, T.; Jiang, H.; Kara, L.B. TopologyGAN: Topology Optimization Using Generative Adversarial Networks Based on Physical Fields Over the Initial Domain. *J. Mech. Des.* **2021**, *143*, 031715. [CrossRef]
41. Maze, F.; Ahmed, F. Diffusion Models Beat GANs on Topology Optimization. In Proceedings of the Thirty-Seventh AAAI Conference on Artificial Intelligence, Washington DC, USA , 7–14 February 2023; Volume 37, pp. 9108–9116.
42. Yu, Y.; Hur, T.; Jung, J.; Jang, I.G. Deep learning for determining a near-optimal topological design without any iteration. *Struct. Multidiscip. Optim.* **2019**, *59*, 787–799. [CrossRef]
43. Zheng, S.; He, Z.; Liu, H. Generating three-dimensional structural topologies via a U-Net convolutional neural network. *Thin-Walled Struct.* **2021**, *159*, 107263. [CrossRef]
44. Chandrasekhar, A.; Suresh, K. TOuNN: Topology Optimization using Neural Networks. *Struct. Multidiscip. Optim.* **2021**, *63*, 1135–1149. [CrossRef]

45. Chen, H.; Joglekar, A.; Kara, L.B. Topology Optimization Using Neural Networks With Conditioning Field Initialization for Improved Efficiency. *J. Mech. Des.* **2024**, *146*, 061702. [CrossRef]
46. Wang, Z.; Melkote, S.; Rosen, D.W. Generative Design by Embedding Topology Optimization into Conditional Generative Adversarial Network. *J. Mech. Des.* **2023**, *145*, 111702. [CrossRef]
47. Li, K.; Ye, W.; Gao, Y. A 3D Structure Mapping-Based Efficient Topology Optimization Framework. *J. Mech. Des.* **2023**, *145*, 081703. [CrossRef]
48. Qian, C.; Ye, W. Accelerating gradient-based topology optimization design with dual-model artificial neural networks. *Struct. Multidiscip. Optim.* **2021**, *63*, 1687–1707. [CrossRef]
49. Yin, J.; Wen, Z.; Li, S.; Zhang, Y.; Wang, H. Dynamically configured physics-informed neural network in topology optimization applications. *Comput. Methods Appl. Mech. Eng.* **2024**, *426*, 117004. [CrossRef]
50. Ho, J.; Jain, A.; Abbeel, P. Denoising diffusion probabilistic models. *Adv. Neural Inf. Process. Syst.* **2020**, *33*, 6840–6851.
51. Nichol, A.; Dhariwal, P. Improved Denoising Diffusion Probabilistic Models. *Int. Conf. Mach. Learn.* **2021**, *139*, 116041.
52. Ronneberger, O.; Fischer, P.; Brox, T. U-Net: Convolutional Networks for Biomedical Image Segmentation. *Med Image Comput. Comput.-Assist. Interv.* **2015**, *9351*, 234–241.

applied sciences

MDPI

Article

Automated Design and Parametric Modeling of Excavator Buckets

Nikolay Stankov * and Aleksandar Ivanov *

Department of Machine Tools and Manufacturing, Faculty of Mechanical and Manufacturing Engineering, University of Ruse, 8 Studentska Street, 7017 Ruse, Bulgaria
* Correspondence: nstankov@uni-ruse.bg (N.S.); akivanov@uni-ruse.bg (A.I.)

Abstract: This paper presents the process of automated designing and parametric modeling of excavator buckets using the CAD system. A methodology has been developed that covers all stages of design, from the creation of 3D models of the parts, sub-assemblies, and assemblies to the generation of design documentation, using template files. The presented approach allows the creation of parametric models with multiple configurations, which cover different variants of the products according to their technical characteristics and purpose. This paper describes in detail the technological process, including the manufacture of parts and the assembling of sub-assemblies, in accordance with modern production requirements. Bucket classification according to excavator operating weight is included. Parametric modeling of the main components, such as the base and upper part of the bucket, is also described. In addition, an algorithm for automation of design documentation has been developed that integrates PDM systems for engineering data management. Using the presented methodology leads to a reduction in the volume of created documentation and optimizes the design process, providing the opportunity for rapid adaptation to various client requirements. Data management within a PDM system ensures centralized storage, actuality, and accessibility of information, which contribute to more efficient production and management of product data.

Keywords: excavator buckets; excavators; methodology; development; 3D models; automated designing; parametric modeling; CAD system; template files; design documentation; PDM system

1. Introduction

This paper introduces the process of developing excavator buckets with a CAD system. The parts, sub-assemblies, and assembly of the bucket are parametrically designed. Different bucket models differ in their purpose, dimensions, and construction. The buckets are mounted on excavators and used for excavation work [1,2].

The bucket is designed and developed with the CAD system SolidWorks 2022 [3]. Using the CAD system significantly facilitates the process of developing the product. The design is performed in the three-dimensional space. Three-dimensional models of the parts, sub-assemblies, and assemblies are developed. The process of creating the design documentation is automated through the use of template files.

2. Literature Review

2.1. Parametric Modeling of Products with a CAD System

Three-dimensional models of parts, sub-assemblies, and assemblies are created through CAD systems. Design documentation is being developed—part drawings, assembly drawings, assembly specifications, and individual product specifications. Three-dimensional models can be used to create new design documentation. In 3D models of the parts and assemblies, the additional information is set by attributes (users and systems ones). The information from the attributes can be used for documentation management with a PDM system, as well as by other departments in a company with an ERP system [3–6].

Parametric modeling is used when the designed products have many versions and varieties. A single 3D model is created, where different versions of the product are presented with configurations. This results in a small amount of design documentation for a specific product. The big amount of documentation can cause many problems, for example, how to manage, organize, classify, and archive the documentation; who has the right to change and update it; the documents should always be with the latest version; and which other departments of the company have access to this documentation and the specialized information they are interested in [5,6].

Parametric modeling reduces the time of the design process and increases its efficiency. Design automation is achieved, which means that designers do not have to redesign the product when there is a change in the dimensions of the model. Engineers can model parametrically dimensions of the model, geometric constructions in the model (Features), constraints between parts in an assembly (Mates), etc. All of these parameters can be adjusted by making a few minor changes. In this way, product variants are created very quickly compared to the traditional modeling approach.

High accuracy is achieved in product modeling. Various algorithms are used in the creation of 3D models, which leads to a reduction in modeling errors.

Parametric modeling creates variations of 3D models and assemblies. It can be easily integrated into various manufacturing processes, which leads to faster product manufacturing.Parametric modeling is used for structural analysis or for analyzing the kinematics of machines. The reason is the ability to quickly process extremely complex geometric data.

2.2. Design Documentation Management Through PDM System, Application of ERP Systems

CAD systems are used to develop a large volume of design documentation—3D models, drawings, specifications, etc. Specialized systems, like PDM and ERP systems, are used to manage and maintain it. To manage the design documentation of the products considered in this publication, the PDM system SolidWorks PDM 2022 and the ERP system RDS ERP are used [3,4].

Product Data Management (PDM) systems are specialized software used to manage product data. The term "product data" means all information related to the product. PDM systems manage arrays of data and information necessary for the design, manufacture, and assembling of the products, as well as for their maintenance. The functionalities of PDM systems have been improved and, in addition to the Design and Engineering Departments, they are now used by other departments such as Supplies, Production, and Maintenance. This means that PDM systems are developed to manage and control the huge volume of electronic data created by CAD/CAM/CAE systems.

Enterprise Resource Planning (ERP) systems are used in all processes necessary for the functioning of a company—people management, finances, production, availability of materials or products, supply system, services, orders, etc. One of the main functions of ERP systems is the shared database, i.e., the employees from different departments operate with the same information for their specific needs.

Both systems are part of the so-called Product Lifecycle Management (PLM). This means that a specific product is tracked from the creation of the product concept to the end of its life cycle.

2.3. Designing Buckets for Excavators

The design of the discussed buckets is a long process, which is implemented in many stages. Each stage has its own specific characteristics that can be presented in the following sequence [7]:

PRELIMINARY PHASE

- Preliminary design—CAD software SolidWorks 2022 is used to model the parts, sub-assemblies, and assemblies of the designed bucket model;
- Simulation inspections—the buckets are inspected with the Finite Element Method;
- Three-dimensional model adjustments—depending on the results obtained during the previous stage, the necessary structural changes are made to the parts, sub-assemblies, and assemblies of the bucket;
- Production of the preliminary prototypes—a real prototype of the bucket is produced;
- Conducting prototype tests—prototype tests are carried out in real conditions.

DESIGN PHASE

- Adjustments to 3D models and drawings—based on the results from the testing stage, the necessary design changes are made to the parts, sub-assemblies, and assemblies of the bucket;
- Simulation inspections—reverification of the bucket using the Finite Element Method;
- Production of the design prototypes—a real prototype of the bucket is produced;
- Conducting prototype tests—prototype tests are carried out in real conditions;
- Design freeze—the design process of the product is finalized.

INDUSTRIALIZATION PHASE

- Final modeling—3D models of the parts, sub-assemblies, and assemblies of the bucket are finalized and drawings and specifications are developed;
- Engineering Pilot—mass production technologies are used to manufacture the bucket;
- Testing and Certification—the produced buckets are tested and certified.

PRODUCT LAUNCH

- Start of production—serial production of the bucket begins.

This publication will only consider 3D modeling of products or modification of already created 3D files by using the capabilities of modern CAD systems. Emphasis will be placed on the capabilities of parametric modeling of buckets and on the automated creation of design documentation (drawings of parts, drawings of assemblies, and specifications). The CAD system used is SolidWorks 2022 [3].

Regardless of the type of CAD system used, the principles of modeling are the same. Popular CAD systems for modeling and simulations in 3D space are Autodesk Inventor [8], Solid Edge [9], Creo [10], CATIA [11], Altair Inspire [12], and others.

The aim of using CAD systems is to increase the efficiency of engineering work by reducing design time and the possibility of errors. This can only be achieved through automation of product design. Another effect is to reduce the volume of design documentation.

2.4. Bucket Classification

Depending on the operational weight of the excavator, a classification of the buckets by classes was made (Table 1). Each bucket, depending on its class, differs in shape, dimensions, length, volume, and other specific indicators [1,2,13–23].

Table 1. Classification of buckets, depending on the operating weight of the excavator.

Excavator Operating Weight, t	**5** **8**	**9** **13**	**14** **16**	**17** **22**	**23** **26**	**27** **30**	**31** **38**	**40** **50**	**55** **65**	**70** **90**	**95** **120**	**130** **150**
Bucket class	A	B	C	D	E	F	G	H	J	K	L	M

Excavator buckets are characterized by the following features, which 3D models must also possess: bucket length; attachment for establishing the arm of the excavator; and type, size, and number of used teeth.

Depending on the class, the length of the buckets is from 600 mm to 1500 mm and, for larger classes, it reaches up to 2000 mm. The excavator arm is attached to the bucket attachment holes using pins. The bucket attachment is different for all the models of excavators. The type and size of the teeth depends on the class of the bucket and the customer requirements, and their number is determined by the length of the bucket.

It should be noted that it is possible to manufacture special bucket designs [1,2,13]. For this purpose, standard designs are used, to which the necessary changes are made. This happens often, since sometimes buckets are used under specific operating conditions that require specialized designs, which are different from standard models. Special bucket designs are shown in Figure 1. These are, for example, buckets with a length of 400 mm and 500 mm, buckets with specialized attachments, for mounting the quick-change devices (Quick Couplers), buckets with holes on the side surfaces, buckets without teeth, etc.

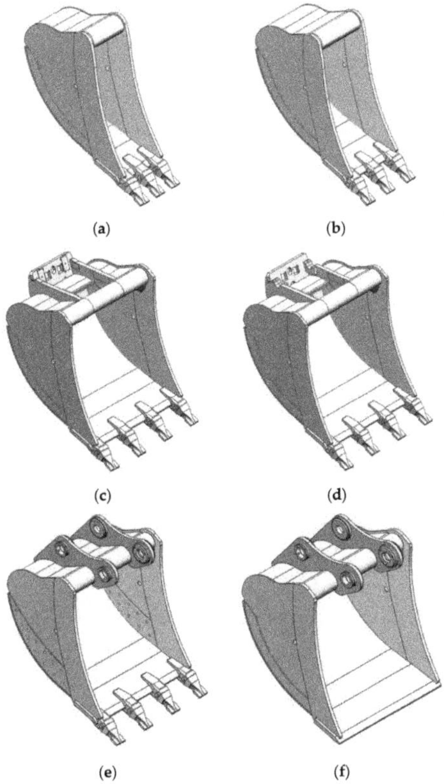

Figure 1. Special bucket designs class D: (**a**) 400 mm; (**b**) 500 mm; (**c**) 1000 mm—QC (Quick Coupler); (**d**) 1000 mm—QC1 (Quick Coupler 1); (**e**) 1000 mm—with holes; (**f**) 1000 mm—without teeth.

According to the classification presented in Table 1, multiple 3D models need to be developed as follows for:

- Each bucket length for each class;
- All types of bucket attachments used by various excavators;
- Each different tooth model.

In conclusion, it can be said that, with so many different versions and variations of bucket designs, the task of designing them becomes very difficult, both in terms of the huge amount of work involved in the design and in terms of the complexity of the work and control of the entire work process.

A certain product can be designed by applying two main methods—traditional modeling or parametric modeling. Both approaches are applied in practice. Depending on the products being developed, the method to be used is also chosen.

In addition, the development of 3D models must be consistent with the technology for manufacturing the buckets, as well as with the possibility of using the models from the PDM and ERP system.

2.5. Bucket Manufacturing Technology

The bucket manufacturing technology includes several stages.

The first stage is manufacturing the upper part of the bucket together with the backet attachment. The parts that are made of sheet metal are cut from sheet metal with the appropriate thickness, after which some of them are machined on CNC machines (Figure 2) [24,25], and others are bent to appropriate shape and size. Other parts are cut from tubes with the appropriate dimensions; then, they are machined on lathes machines.

Figure 2. Machining of the ear part from the backet attachment of the upper part.

All the parts are assembled with a fixture, tack welded, and welded. The upper part is machined on a machining center; the mounting holes are machined (Figure 3) [24,25]. Specialized CAM software is used to create the machining program—SolidWorks CAM 2022 [3].

The second stage is assembling the entire bucket using a fixture. The upper part is installed on the fixture and all other parts of the bucket are added. The parts are tack welded. The bucket is removed from the fixture and welded. After welding the bucket, the teeth are assembled and welded. The last stage is the control of the dimensions of the bucket. The integrity and quality of the weld beads are checked.

Figure 3. Machining of the mounting holes of the upper part.

3. Parametric Modeling of Excavator Buckets

This publication presents an approach to improve the efficiency of the product design process by using parametric modeling. The CAD system SolidWorks 2022 was used [3].

The SolidWorks CAD system enables parametric modeling through parametric tables or through configurations.

The capabilities of MS Excel are used in the modeling using parametric tables. Parts, sub-assemblies, and assemblies are configured using tables in which values of the parameters are set. Formulas can be used and calculations can be made.

In configuration modeling, configurations are created in parts, sub-assemblies, and assemblies. The values of the parameters are set in the configurations. The publication discusses the method of parametric modeling through using configurations. For each of the bucket classes, a single 3D model has been developed that is parametrically modeled.

To illustrate parametric modeling, a class D bucket will be considered. This bucket class is for excavators from 17 to 22 t, which are widely used in practice.

Figure 4 shows the 3D model of the bucket considered in the publication.

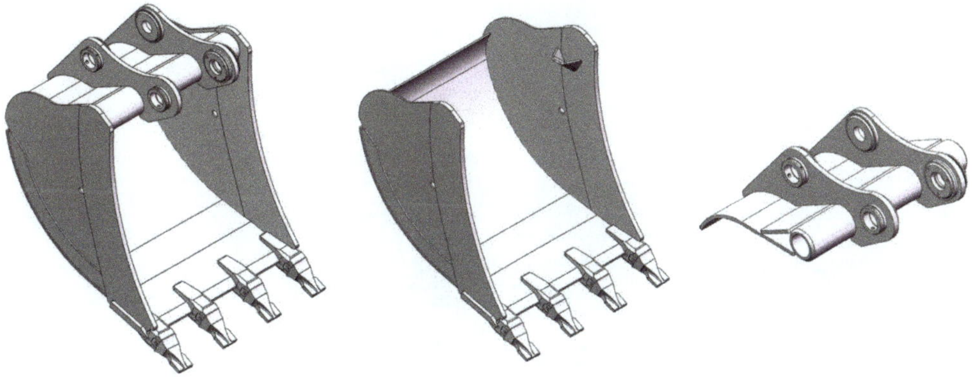

(a) (b) (c)

Figure 4. Excavator bucket class D: (**a**) 3D model of the bucket; (**b**) base part; (**c**) upper part.

The 3D model is composed of two sub-assemblies—a base part, including parts of the bucket and teeth, and an upper part, including all other parts and the backet attachment. Regardless of the bucket class, every 3D bucket model consists of these two basic sub-assemblies.

Each of the parts from the sub-assemblies is developed by using different 3D modeling commands—Extruded Boss/Base, Revolved Boss/Base, Extruded Cut, Revolved Cut, etc.

Parts that are made from sheet metal are created using the module Sheet Metal; in this way, their unfolding is obtained. The parts are assembled into individual sub-assemblies, which are assembled into the assembly of the bucket [13].

When products are developed and modeled with a CAD system, it is appropriate to use a specific design methodology. This makes the modeling process easier and reduces the time required for designing the elements. Figure 5 shows a summarized algorithm of the methodology for designing excavator buckets.

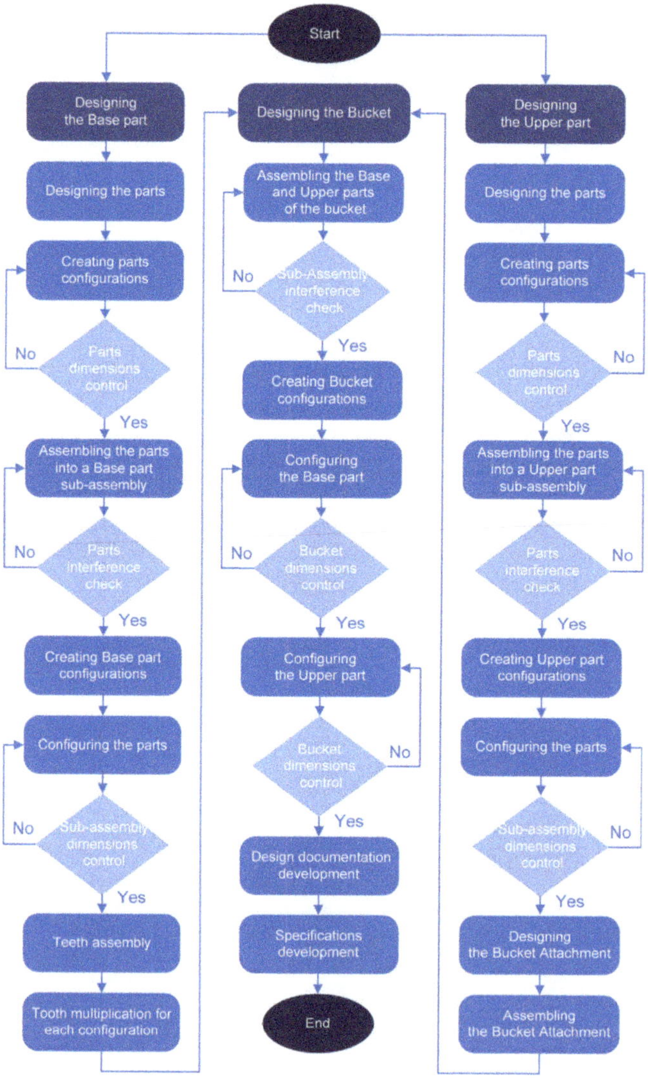

Figure 5. Excavator bucket design algorithm.

3.1. Parametric Modeling of Base Part

Figure 6 shows the algorithm used to design the base part.

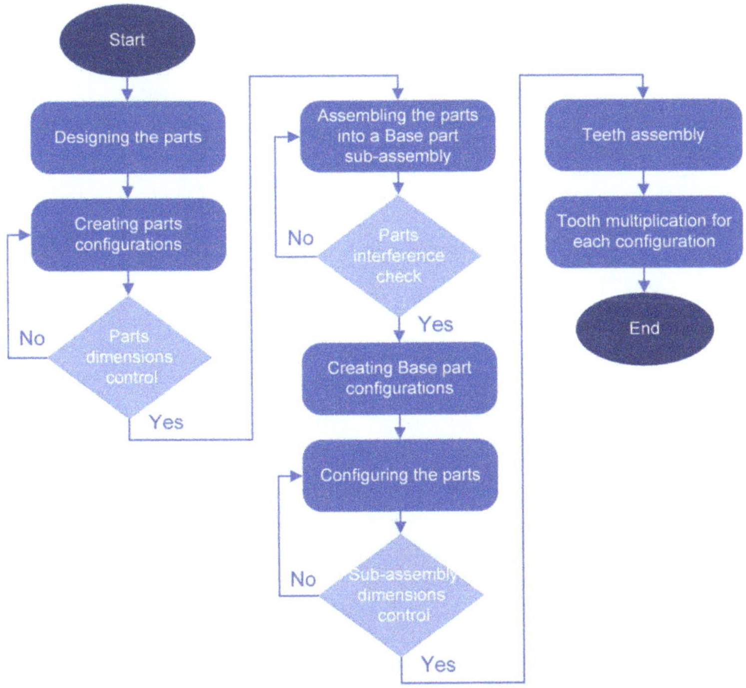

Figure 6. Base part design algorithm.

The 3D model of the base part is parametrically modeled through configurations and Figure 7 shows its components.

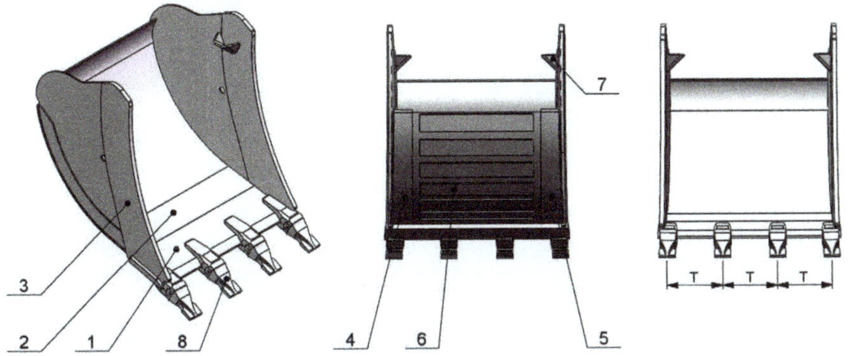

Figure 7. Components of the base part. 1—Blade, 2—Bottom, 3—Side, 4—Left Rib, 5—Right Rib, 6—Horizontal Rib, 7—Internal Rib, 8—Teeth.

Parts with positions 1, 2, and 6 are parametrically modeled and they have different length configurations. This sub-assembly contains three parts that are structurally identical but differ in length.

To model parametrically the 3D model of the base part, configurations are created in the Configuration Manager (Figure 8). Each configuration represents a different length of the base part. Base parts differ in length—from 500 mm (NGPD50FEI) to 1500 mm (NGPD15FEI).

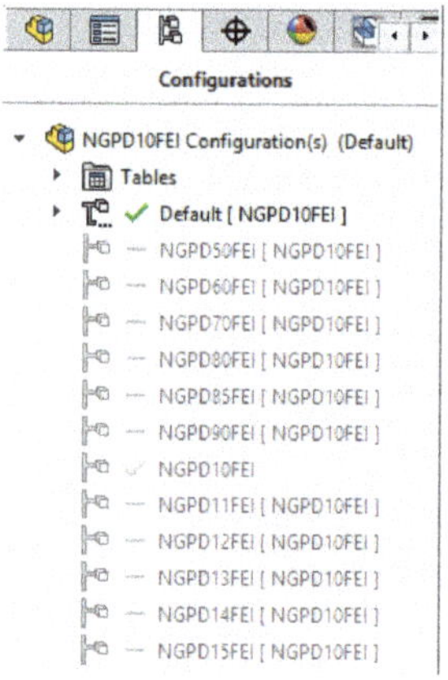

Figure 8. Configurations in the 3D model of the base part.

The next step is to configure the parts in the assembly of the base part. Parts with positions 1 (Blade), 2 (Bottom), and 6 (Horizontal Rib) contain configurations with different lengths. For each configuration of the base part, the corresponding configuration of the parts is set.

Figure 9 shows an example of the configurations of the base part and the specified configurations of the blade part. For the respective lengths of the base part, the blade part has the following configurations—from 500 mm (12L5027035FEP) to 1500 mm (12L1527035FEP).

Configuration Name	12LXX27035FEP-1@NGPD10FEI		
	Suppress	Configuration	Fixed
Default	☐	Default	☐
NGPD50FEI	☐	12L5027035FEP	☐
NGPD60FEI	☐	12L6027035FEP	☐
NGPD70FEI	☐	12L7027035FEP	☐
NGPD80FEI	☐	12L8027035FEP	☐
NGPD85FEI	☐	12L8527035FEP	☐
NGPD90FEI	☐	12L9027035FEP	☐
NGPD10FEI	☐	12L1027035FEP	☐
NGPD11FEI	☐	12L1127035FEP	☐
NGPD12FEI	☐	12L1227035FEP	☐
NGPD13FEI	☐	12L1327035FEP	☐
NGPD14FEI	☐	12L1427035FEP	☐
NGPD15FEI	☐	12L1527035FEP	☐
< Creates a new configuration. >			

Figure 9. Configuring the blade part.

A single 3D model of the blade part has been developed. Configuring the blade part involves setting different values for its length in the corresponding configuration (Table 2). In this case, parametric modeling is implemented by configuring a dimension from the 3D model of the part.

If the part is developed using the traditional modeling approach, a different 3D model of the blade part with the corresponding length must be modeled for each bucket model. In the case of class D buckets, 12 different 3D models of the part must be designed. Modeling multiple parts is a prerequisite for making errors. In addition, the modeling time increases.

The number of teeth is different for different bucket configurations. Each tooth consists of several parts (Figure 10a). In the 3D model of the bucket, the parts are assembled and, using the Linear Component Pattern function, multiplied to the required number. For each configuration, the corresponding number of teeth is multiplied by the function. Depending on the configuration that is used, the corresponding geometric construction (Feature) is activated, while the others are suppressed using the Suppressed command (Figure 10b). In this case, parametric modeling is implemented by configuring a geometric construction (Feature) from the 3D model of the bucket.

Table 2. Configurations of the blade part.

№	Bucket Size, mm	Base Part Configuration	Part Configuration	Part Length, mm
1	500	NGPD50FEI	12L5027035FEP	512
2	600	NGPD60FEI	12L6027035FEP	612
3	700	NGPD70FEI	12L7027035FEP	712
4	800	NGPD80FEI	12L8027035FEP	812
5	850	NGPD85FEI	12L8527035FEP	862
6	900	NGPD90FEI	12L9027035FEP	912
7	1000	NGPD10FEI	12L1027035FEP	1012
8	1100	NGPD11FEI	12L1127035FEP	1112
9	1200	NGPD12FEI	12L1227035FEP	1212
10	1300	NGPD13FEI	12L1327035FEP	1312
11	1400	NGPD14FEI	12L1427035FEP	1412
12	1500	NGPD15FEI	12L1527035FEP	1512

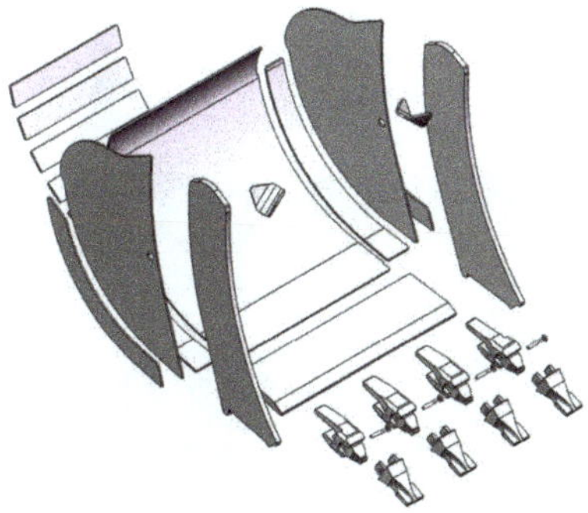

(a)

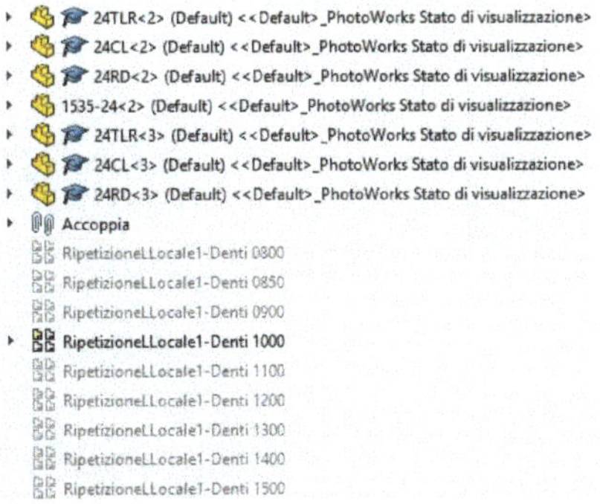

(b)

Figure 10. Configuration of teeth for a 1000 mm bucket: (**a**) Explode view of the 3D model of the base part; (**b**) tooth configuration.

There are different types of bucket teeth and, therefore, different sizes, for example, FK Fast Key, FS Orca S, FE Orca, Super V, CO CAT J, CAT K, KM Komatsu, VL Volvo, etc. [1,2,13,18]. If a bucket with a different tooth model has to be developed using the traditional modeling approach, a completely new 3D model of the bucket must be modeled too.

The following figure shows the tooth configuration (FK Fast Key) for a class D bucket, with a length of 1000 mm.

Figure 11 shows different lengths of the base part, parametrically modeled with configurations.

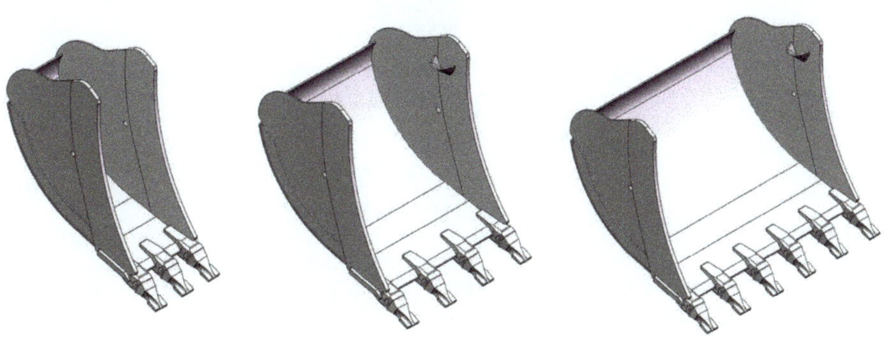

(a) (b) (c)

Figure 11. Parametric models of the base part: (**a**) 500 mm; (**b**) 1000 mm; (**c**) 1500 mm.

3.2. Parametric Modeling of Upper Part

Figure 12 shows the algorithm used to design the upper part.

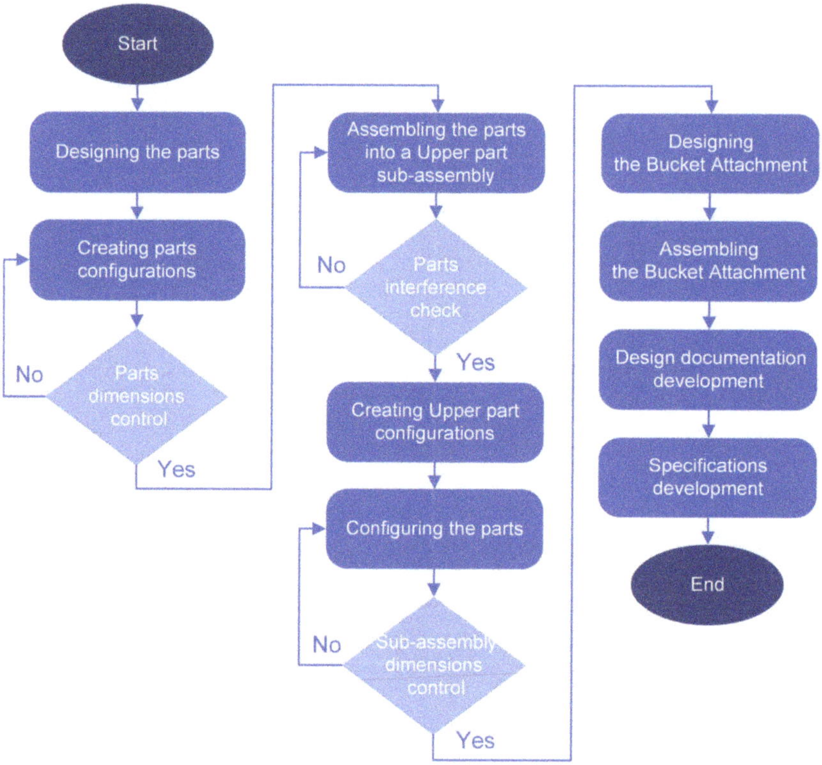

Figure 12. Upper part design algorithm.

The 3D model of the upper part is parametrically modeled through configurations and Figure 13 shows its components.

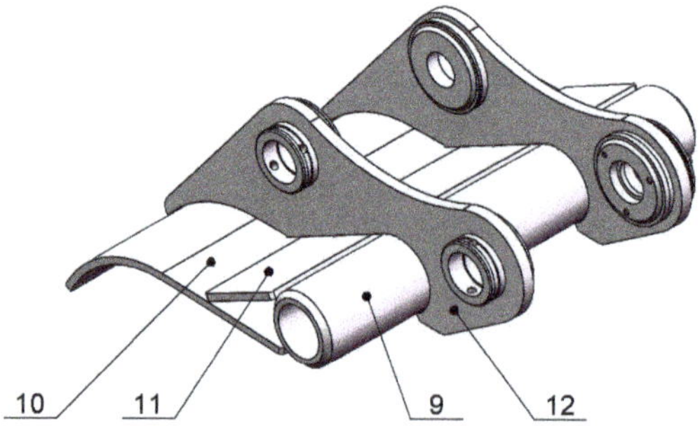

Figure 13. Components of the upper part. 9—Tube, 10—Cover, 11—Rib, 12—Bucket Attachment.

Parts with positions 9, 10, and 11 are parametrically modeled and they have different length configurations. This sub-assembly also contains three parts that are structurally identical but differ in length. The Bucket Attachment 12 is the transitional link between the bucket and the excavator arm. The bucket attachment is different for different excavator models.

To model parametrically the 3D model of the upper part, configurations are created in the Configuration Manager (Figure 14). Each configuration represents a different length of the upper part. Upper parts differ in length—from 600 mm (GPD6025-B33) to 1500 mm (GPD1525-B33).

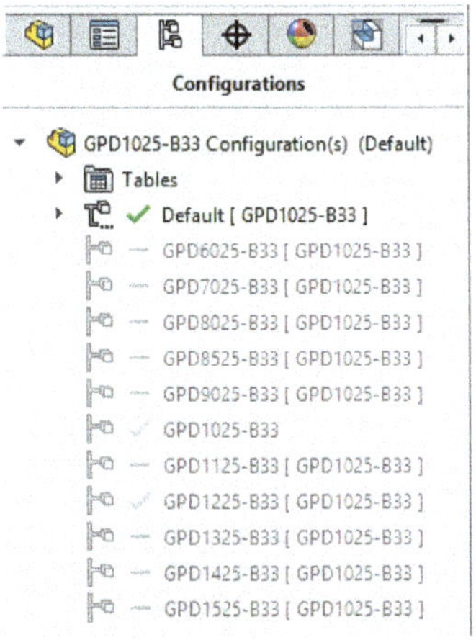

Figure 14. Configurations in the 3D model of the upper part.

The next step is to configure the parts in the assembly of the upper part. Parts with positions 9 (Tube), 10 (Cover), and 11 (Rib) contain configurations with different lengths. For each configuration of the upper part, the corresponding configuration of the parts is set.

Figure 15 shows an example of the configurations of the upper part and the specified configurations of the tube part. For the corresponding lengths of the upper part, the tube part has the following configurations—from 600 mm (D6070) to 1500 mm (D1570).

Modify Configurations

Configuration Name	DXX70-1@GPD1026-B33		
	Suppress	Configuration	Fixed
Default	☐	Default	☐
GPD6025-B33	☐	D6070	☐
GPD7025-B33	☐	D7070	☐
GPD8025-B33	☐	D8070	☐
GPD8525-B33	☐	D8570	☐
GPD9025-B33	☐	D9070	☐
GPD1025-B33	☐	D1070	☐
GPD1125-B33	☐	D1170	☐
GPD1225-B33	☐	D1270	☐
GPD1325-B33	☐	D1370	☐
GPD1425-B33	☐	D1470	☐
GPD1525-B33	☐	D1570	☐
< Creates a new configuration. >			

Figure 15. Configuring the tube part.

A single 3D model of the tube part has been developed. Configuring the tube part involves setting different values for its length in the corresponding configuration (Table 3). Parametric modeling is implemented by configuring a dimension from the 3D model of the part.

If the part is developed using the traditional modeling approach, a different 3D model of the tube part with the corresponding length must be modeled for each bucket model. In the case of class D buckets, 11 different 3D models of the part must be designed. Modeling multiple parts increases modeling time and the possibility of making errors.

Figure 16 shows different lengths of the upper part, parametrically modeled with configurations.

Table 3. Configurations of the tube part.

№	Bucket Size, mm	Upper Part Configuration	Part Configuration	Part Length, mm
1	600	GPD6025-B33	D6070	540
2	700	GPD7025-B33	D7070	640
3	800	GPD8025-B33	D8070	740
4	850	GPD8525-B33	D8570	790
5	900	GPD9025-B33	D9070	840
6	1000	GPD1025-B33	D1070	940
7	1100	GPD1125-B33	D1170	1040
8	1200	GPD1225-B33	D1270	1140
9	1300	GPD1325-B33	D1370	1240
10	1400	GPD1425-B33	D1470	1340
11	1500	GPD1525-B33	D1570	1440

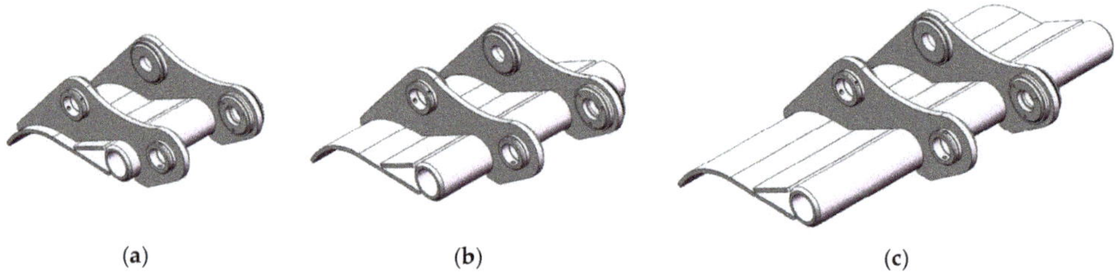

Figure 16. Parametric models of the upper part: (**a**) 600 mm; (**b**) 1000 mm; (**c**) 1500 mm.

3.3. Parametric Modeling of Buckets

Figure 17 shows the algorithm used to design the bucket.

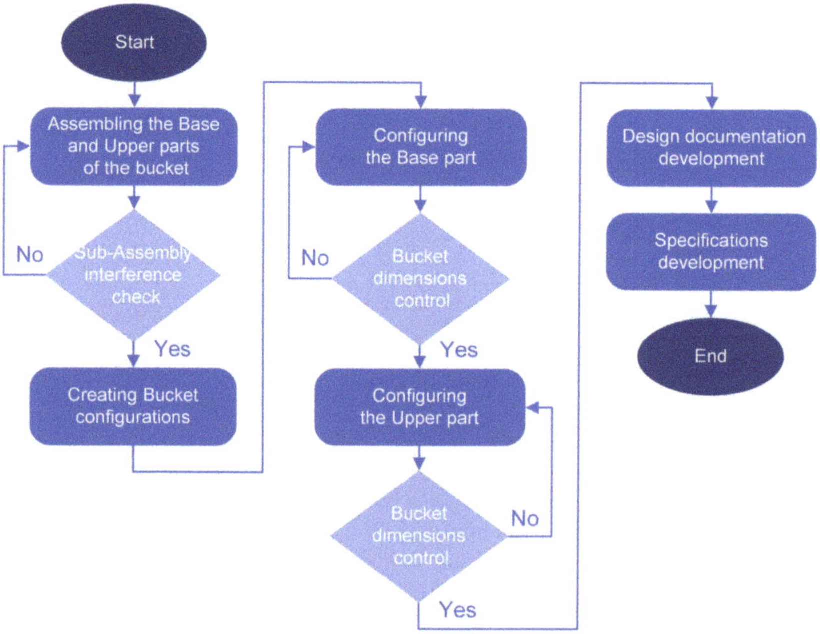

Figure 17. Bucket design algorithm.

The 3D model of each bucket is parametrically modeled and has configurations with different lengths. The number of teeth is configured for each length. Depending on the model of excavator for which the respective bucket is designed, the bucket attachment is different, as well as the type and size of the teeth.

The 3D bucket model consists of the sub-assemblies—the base part and upper part are assembled using the main planes of the Assembly, i.e., Front Plane (Piano Frontale), Top Plane (Piano Superiore), and Right Plane (Piano Destro) (Figure 18).

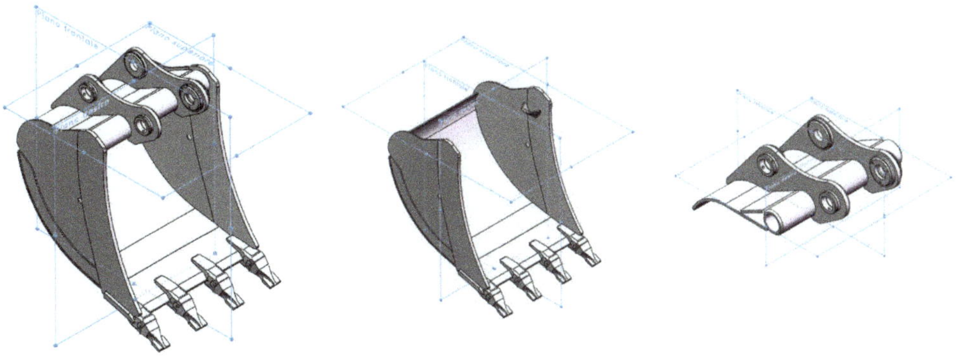

(**a**) (**b**) (**c**)

Figure 18. Assembling the basic sub-assemblies of the 3D model of the bucket: (**a**) 3D model of the bucket; (**b**) base part; (**c**) upper part.

To perform assembly, the 3D models of the sub-assemblies have in the same place co-ordinate systems and planes that match with the co-ordinate system Origin and the base planes Front Plane, Top Plane, and Right Plane of the 3D model of the bucket. The 3D model that is assembled in this way allows quick replacement of the base or upper part with other similar ones through the function Replace Components. This allows automated generation of new 3D models of buckets, with different base parts or upper parts. For example, all special bucket designs shown in Figure 2 are modeled in this way.

To model parametrically the 3D model of the bucket, configurations are created in the Configuration Manager (Figure 19). Each configuration represents a different length of the bucket. Buckets differ in length—from 600 mm (NGPD60FE-B33) to 1500 mm (NGPD15FE-B33).

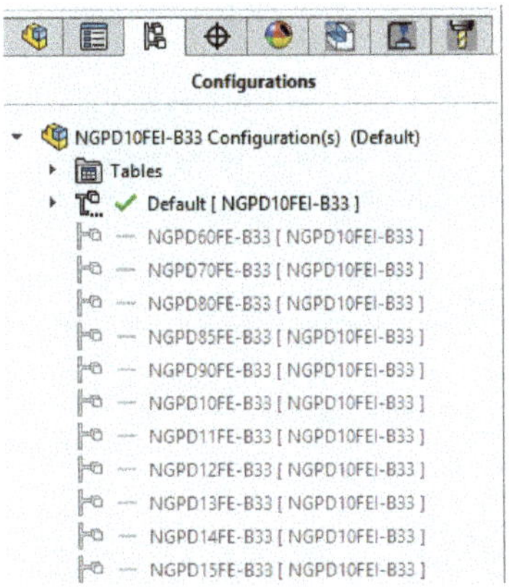

Figure 19. Configurations in the 3D model of the bucket.

The next step is to configure the sub-assemblies in the assembly of the bucket. The base part and the upper part also contain configurations. For each bucket configuration, the corresponding configuration from the specific sub-assemblies is set. In this case, parametric modeling is implemented by configuring the sub-assemblies of the 3D model of the bucket.

Figures 20 and 21 show examples of bucket configurations and configurations that are set for the base part and the upper part.

Modify Configurations

Configuration Name	NGPD10FEI-1@NGPD10FEI-B33		
	Suppress	Configuration	Fixed
Default	☐	Default	☐
NGPD60FE-B33	☐	NGPD60FEI	☐
NGPD70FE-B33	☐	NGPD70FEI	☐
NGPD80FE-B33	☐	NGPD80FEI	☐
NGPD85FE-B33	☐	NGPD85FEI	☐
NGPD90FE-B33	☐	NGPD90FEI	☐
NGPD10FE-B33	☐	NGPD10FEI	☐
NGPD11FE-B33	☐	NGPD11FEI	☐
NGPD12FE-B33	☐	NGPD12FEI	☐
NGPD13FE-B33	☐	NGPD13FEI	☐
NGPD14FE-B33	☐	NGPD14FEI	☐
NGPD15FE-B33	☐	NGPD15FEI	☐
< Creates a new configuration. >			

Figure 20. Configuring the base part.

Modify Configurations

Configuration Name	GPD1025-B33-1@NGPD10FEI-B33		
	Suppress	Configuration	Fixed
Default	☐	Default	☐
NGPD60FE-B33	☐	GPD6025-B33	☐
NGPD70FE-B33	☐	GPD7025-B33	☐
NGPD80FE-B33	☐	GPD8025-B33	☐
NGPD85FE-B33	☐	GPD8525-B33	☐
NGPD90FE-B33	☐	GPD9025-B33	☐
NGPD10FE-B33	☐	GPD1025-B33	☐
NGPD11FE-B33	☐	GPD1125-B33	☐
NGPD12FE-B33	☐	GPD1225-B33	☐
NGPD13FE-B33	☐	GPD1325-B33	☐
NGPD14FE-B33	☐	GPD1425-B33	☐
NGPD15FE-B33	☐	GPD1525-B33	☐
< Creates a new configuration. >			

Figure 21. Configuring the upper part.

Figure 22 shows different bucket lengths parametrically modeled with configurations.

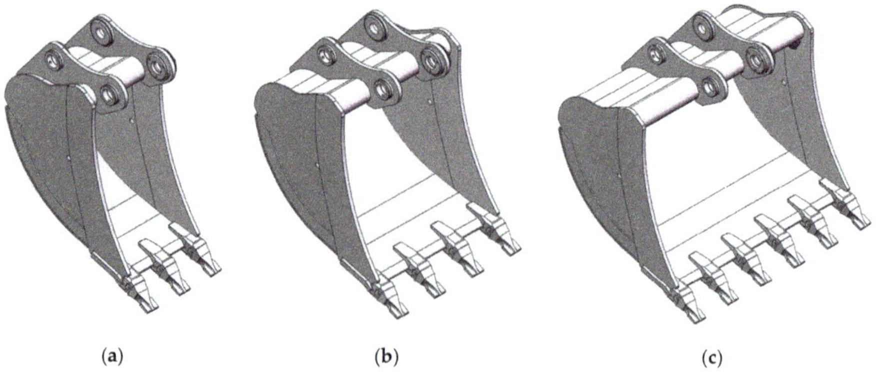

(**a**) (**b**) (**c**)

Figure 22. Parametric models of the bucket: (**a**) 600 mm; (**b**) 1000 mm; (**c**) 1500 mm.

3.4. Design Documentation of the Buckets

Complete design documentation has been made for the developed excavator buckets. The process of its creation is automated by using document Template files in the SolidWorks CAD system.

Template files for the Part and Assembly documents are used to create the 3D models of the parts and assemblies. In the template of each 3D model of the part or assembly, from the File Properties menu, in the Summary Information window, in the Custom section, the required attributes (users and systems ones) are selected. The attributes for each configuration are filled in the Configuration Specific window [5,6].

After the attributes of the 3D models have been filled in, the design documentation of the product (drawings and specifications) is prepared. For the considered example of an excavator bucket, drawings of all parts, the upper part sub-assembly, and the assembly of the bucket are developed.

For the development of the drawings, template files are used for the Drawing documents, which represent the standard drawing formats—A4, A3, A2, A1, and A0. Information from the attributes is automatically visualized in the drawing tables, in the assembly drawing specifications, and in individual product specifications [5,6].

Figure 23 shows examples of completed templates for the tube part and the sub-assembly of the upper part.

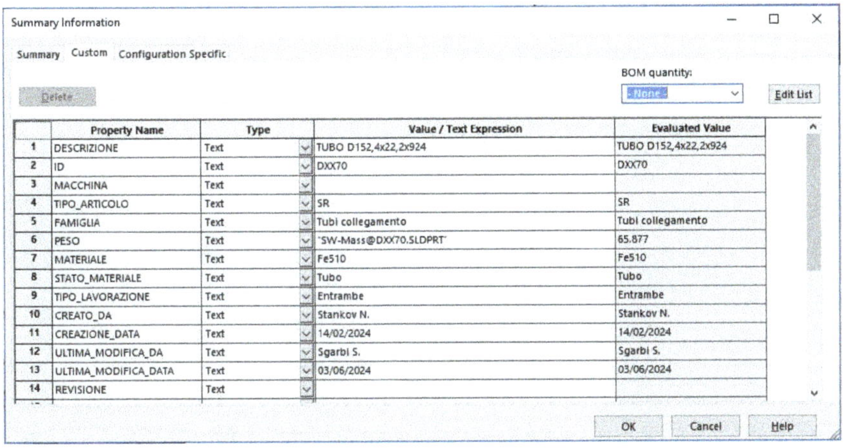

	Property Name	Type	Value / Text Expression	Evaluated Value
1	DESCRIZIONE	Text	TUBO D152,4x22,2x924	TUBO D152,4x22,2x924
2	ID	Text	DXX70	DXX70
3	MACCHINA	Text		
4	TIPO_ARTICOLO	Text	SR	SR
5	FAMIGLIA	Text	Tubi collegamento	Tubi collegamento
6	PESO	Text	"SW-Mass@DXX70.SLDPRT"	65.877
7	MATERIALE	Text	Fe510	Fe510
8	STATO_MATERIALE	Text	Tubo	Tubo
9	TIPO_LAVORAZIONE	Text	Entrambe	Entrambe
10	CREATO_DA	Text	Stankov N.	Stankov N.
11	CREAZIONE_DATA	Text	14/02/2024	14/02/2024
12	ULTIMA_MODIFICA_DA	Text	Sgarbi S.	Sgarbi S.
13	ULTIMA_MODIFICA_DATA	Text	03/06/2024	03/06/2024
14	REVISIONE	Text		

Figure 23. Template of the tube part.

The following attributes are used in the template of each Part—Descrizione (name of the part), ID (drawing number), Macchina (assembly from which the part is taken), Peso (weight), Materiale (designation and type of the material from which it is made), Tipo Lavorazione (manufacturing operations), Creato Da (developed the document), Ultima Modifica Da (checked the document), Revisione (revision of the document), etc.

Figure 24 shows the drawing of a tube part, with an automatically filled table, based on the attributes specified in the 3D model of the part.

If this drawing is made using the traditional modeling approach, the drawing table should be filled in manually. Apparently, there would be errors as a result of repeated, identical actions by the designer.

Figure 25 shows the attributes are used in the template of each Assembly—Descrizione (name of the assembly), ID (drawing number), Macchina (assembly on which the current assembly is based), Peso (weight), Tipo Lavorazione (manufacturing operations), Creato Da (developed the document), Ultima Modifica Da (checked the document), Revisione (revision of the document), etc.

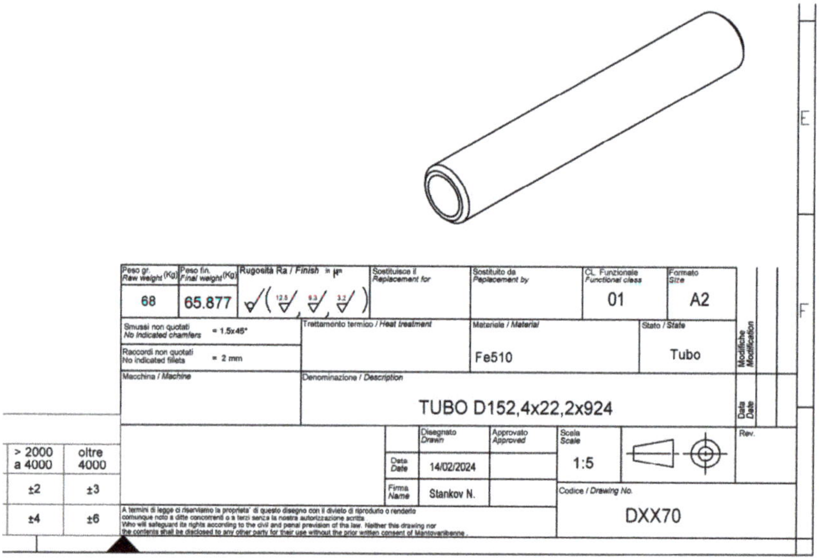

Figure 24. Drawing of the tube part.

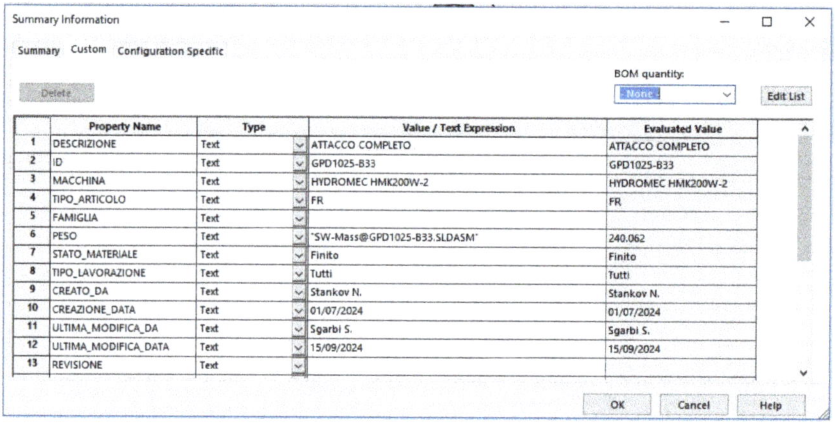

	Property Name	Type	Value / Text Expression	Evaluated Value
1	DESCRIZIONE	Text	ATTACCO COMPLETO	ATTACCO COMPLETO
2	ID	Text	GPD1025-B33	GPD1025-B33
3	MACCHINA	Text	HYDROMEC HMK200W-2	HYDROMEC HMK200W-2
4	TIPO_ARTICOLO	Text	FR	FR
5	FAMIGLIA	Text		
6	PESO	Text	"SW-Mass@GPD1025-B33.SLDASM"	240.062
7	STATO_MATERIALE	Text	Finito	Finito
8	TIPO_LAVORAZIONE	Text	Tutti	Tutti
9	CREATO_DA	Text	Stankov N.	Stankov N.
10	CREAZIONE_DATA	Text	01/07/2024	01/07/2024
11	ULTIMA_MODIFICA_DA	Text	Sgarbi S.	Sgarbi S.
12	ULTIMA_MODIFICA_DATA	Text	15/09/2024	15/09/2024
13	REVISIONE	Text		

Figure 25. Template of the upper part.

Figure 26 shows a drawing of the upper part, with an automatically filled table and specification, based on the attributes set in the 3D model of the assembly.

In the traditional modeling approach, the table on the drawing of the assembly is filled in manually. The specification is created in the drawing or made as a separate document and, again, filled in manually. Many mistakes are made and the time for creating increases significantly.

After the 3D models of the parts, sub-assemblies, assemblies, and all drawings of the buckets have been developed, the next step is their management by the specialized systems—PDM and ERP. The design documentation is managed by the PDM system SolidWorks PDM 2022 [3]. The various departments of the company have access to this information and operate with it through the ERP system RDS ERP [4].

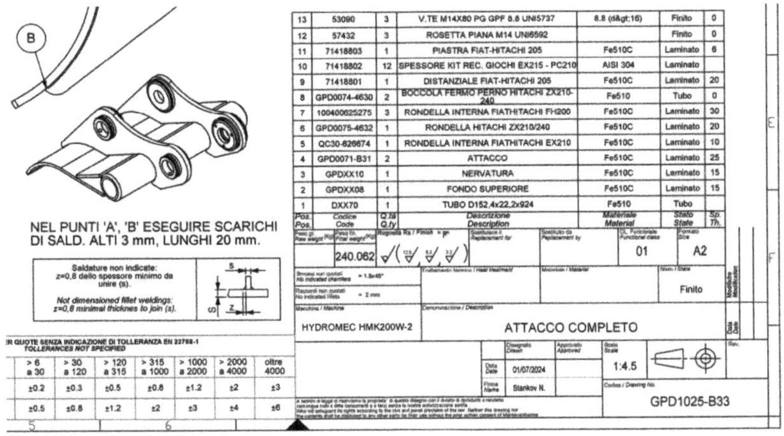

Figure 26. Drawing of the upper part.

In the PDM system, each configuration of the 3D model of a part or assembly is entered as a separate product. The system automatically fills in the fields on the Product Card based on the information set in the attributes of the part or the assembly for the corresponding configuration. If it is necessary in the product card, additional information can be added manually.

PDM systems are used when there is a big amount of design, technological, production, and other types of documentation accompanying the products that are developed, manufactured, and maintained. PDM systems solve the problems related to the maintenance, storage, and actuality of the large volume of documentation connected to the products [5,6].

4. Bucket Modeling Methods

A certain product can be designed by using traditional modeling or configurational parametric modeling. Both approaches are applied in practice.

The parametric modeling approach through configurations has been chosen because of the advantages it has. The reason for using this approach becomes clear after comparing the two modeling methods.

4.1. Traditional Bucket Modeling

In the traditional modeling approach, a 3D model is created for each bucket size. The 3D model of the bucket is an Assembly composed of parts and sub-assemblies. All parts are modeled in advance. Parts are structurally the same but differ in length, for different bucket models must be modeled as separate parts. In this bucket model, there are six parts. Some of the parts are the same for all buckets and they are modeled once. Depending on the bucket size, the number of these parts is different. For example, some buckets have four teeth but there are buckets that have fewer, more, or no teeth at all. After modeling the parts, they are assembled into the 3D model of the bucket. Table 4 shows a class D bucket and its components.

This approach has many disadvantages:

- A large number of parts that must be modeled in advance;
- A large number of assemblies to model;
- Possible modeling errors due to the large variety of 3D models of parts and assemblies;
- Creation of a big amount of design documentation—drawings of parts, drawings of assemblies, and specifications;

- Redesigning the product when the size is different.

Table 4. Classification of class D buckets, designed using the traditional modeling approach.

Bucket Size, mm	600	700	800	850	900	1000	1100	1200	1300	1400	1500
Number of unique parts, that do not require changes	29 (28–600 mm, 700 mm)										
Number of parts that require changes	6	6	6	6	6	6	6	6	6	6	6
Total number of parts	61	61	67	67	67	67	71	71	71	71	75
3D model of the bucket (Assembly)	Yes	Yes	Yes	Yes	Yes	Yes	Yes	Yes	Yes	Yes	Yes

All these lead to a decreased efficiency of engineering work during design and an increased possibility of errors.

4.2. Parametric Bucket Modeling

In parametric modeling through configurations, a single 3D model of the bucket is modeled and it includes all its variations (Table 5). The 3D model of the bucket is composed of parts and sub-assemblies. All parts are modeled in advance. Parts that differ only in length are modeled parametrically through configurations. The parts are assembled into sub-assemblies, which are also modeled parametrically. The sub-assemblies are also assembled into the assembly of the bucket, which is also modeled parametrically.

Table 5. Classification of class D buckets, designed using the parametric modeling approach.

Bucket Size, mm	600	700	800	850	900	1000	1100	1200	1300	1400	1500
Number of unique parts, that do not require changes	29 (28–600 mm, 700 mm)										
Number of parts that require changes	6										
Total number of parts	61	61	67	67	67	67	71	71	71	71	75
3D model of the bucket (Assembly)	A single 3D model for all buckets class D										

Parametric modeling through configurations achieves high design efficiency. Design time is significantly reduced, as well as the possibility of making errors in modeling. The volume of design documentation accompanying the product is significantly reduced too. All this leads to an increase in the efficiency of engineering work when designing with CAD systems.

5. Research Methodology and Experimental Results

Several examples are considered to validate the proposed parametric modeling approach with configurations.

First, the traditional modeling approach will be considered. All parts and assembly of the class D buckets will be modeled. The time required to develop the 3D models will be calculated.

Second, the parametric modeling approach will be considered. The same parts and single assembly of the class D bucket will be modeled. The time required to develop them will be calculated.

All design time measurements were performed with the participation of engineers from the university R&D department working on real industrial projects. The projects are performed for external corporate clients. The measured values reflect the time required to design real products ordered by industrial partners who use the R&D department to design their products. This ensures that the results are representative of industrial conditions and validates the effectiveness of the proposed parametric methodology.

5.1. Traditional Modeling of Parts and Assemblies

The modeling of all the parts for the buckets, class D, will be considered. According to the information from Table 4, 29 unique parts and 6 parts need to be modeled for each 3D bucket model. These six parts differ only in length. The total number of parts that need to be modeled is 95 (Table 6).

Table 6. Time for traditional modeling of all bucket parts.

Bucket Size, mm	600	700	800	850	900	1000	1100	1200	1300	1400	1500	Total, h
Number of Unique Parts, that do not Require Changes	29											
Number of parts that require changes	6	6	6	6	6	6	6	6	6	6	6	
Designer I Modeling time, h	32	3	3	3	3	3	3	3	3	3	3	62
Designer II Modeling time, h	33	4	4	4	4	4	4	4	4	4	4	73
Designer III Modeling time, h	31	3	3	3	3	3	3	3	3	3	3	61
Average time, h												65

The 3D models of all parts are modeled in advance. Therefore, in the column for a bucket, size 600 mm, the total time for modeling 29 unique parts is recorded, plus the time for modeling six parts, which are only for this bucket model. In the following columns, the time for modeling the six parts for the corresponding bucket model is recorded. The parts differ only in length.

The modeling of all buckets, class D, will be considered. According to the information in Table 4, 11 3D bucket models need to be modeled (Table 7).

Table 7. Time for traditional modeling of all bucket assemblies.

Bucket Size, mm	600	700	800	850	900	1000	1100	1200	1300	1400	1500	Total, h
Designer I Modeling time, h	8	1	1	1	1	1	1	1	1	1	1	18
Designer II Modeling time, h	9	1.5	1.5	1.5	1.5	1.5	1.5	1.5	1.5	1.5	1.5	24
Designer III Modeling time, h	8	1.5	1.5	1.5	1.5	1.5	1.5	1.5	1.5	1.5	1.5	23
Average time, h												22

First, the 3D model of a bucket, size 600 mm, is modeled. All parts are assembled in an Assembly file. The total time for the complete assembly of the bucket is recorded in the column. The following columns show the time needed for assembling only the six parts that differ in length. The following 3D models of the buckets are created from the base model 600 mm, replacing only the six different parts. The Replace Components function is used. This allows automated generation of new 3D models.

5.2. Parametric Modeling of Parts and Assemblies

The modeling of all the parts for the buckets, class D, will be considered. According to the information from Table 5, 29 unique parts and 6 parts that differ only in length need to be modeled. These six parts will be modeled as a single 3D model with configurations for the corresponding lengths. The total number of parts that need to be modeled is only 35 (Table 8).

Table 8. Time for parametric modeling of all bucket parts.

Bucket Size, mm	600	700	800	850	900	1000	1100	1200	1300	1400	1500	Total, h
Number of Unique Parts, that do not Require Changes						29						
Number of Parts that Require Changes						6						
Designer I Modeling time, h						34						34
Designer II Modeling time, h						35						35
Designer III Modeling time, h						33						33
Average time, h												34

The modeling of all buckets, class D, will be considered. According to the information from Table 5, only one 3D bucket model will be modeled (Table 9).

Table 9. Time for parametric modeling of all bucket assemblies.

Bucket size, mm	600	700	800	850	900	1000	1100	1200	1300	1400	1500	Total, h
Designer I Modeling time, h						10						10
Designer II Modeling time, h						11						11
Designer III Modeling time, h						10						10
Average time, h												10

In parametric modeling, a single 3D model of a bucket is modeled, with configurations for the different lengths. All parts are assembled in an assembly file. For the corresponding bucket configurations, the six parts, which are of different lengths, are configured.

5.3. Errors Occurring in the Modeling

When modeling parts and assemblies with CAD systems, errors of various kinds may occur. Such errors can be:

- Errors in determining the dimensions during the process of designing the parts;
- Errors in 3D modeling of parts;
- Errors in assembling the parts into the bucket assemblies;
- Errors when creating configurations of parts and assemblies;
- Errors in configuring parts and sub-assemblies in assemblies of the buckets;
- Errors when filling in the Template files of the 3D models of the parts and the assemblies;
- Errors when creating drawings of parts and assemblies;
- Errors when filling in the drawing tables;
- Errors when creating and filling in specifications.

In Table 10, the summarized errors in the designing and modeling of excavator buckets are given, using the two approaches—traditional modeling and parametric modeling.

Table 10. Bucket modeling errors.

Modeling Method	Traditional Modeling	Parametric Modeling
Designer I Modeling errors	20	5
Designer II Modeling errors	24	8
Designer III Modeling errors	22	5
Average number of errors	22	6

5.4. Research Results

According to the obtained results for the modeling of the parts and assemblies of the buckets, the following conclusions can be made:

- In traditional modeling, the total number of parts and buckets that need to be modeled is bigger—95 parts and 11 buckets, compared to 35 parts and 1 bucket in parametric modeling;
- In traditional modeling, the average time for modeling the total number of parts and buckets is longer—65 h for parts and 22 h for buckets, compared to 34 h for parts and 10 h for buckets in parametric modeling;
- In traditional modeling, the average number of errors that occur is higher—22, compared to 6 in parametric modeling;
- Parametric modeling significantly increases the efficiency of engineering work—the volume of design documentation created is reduced and the time required for modeling products is reduced;
- As a result of the smaller amount of design documentation, the number of errors in product modeling is significantly reduced;
- Reducing design time is especially important in modern working conditions, where engineering departments consist of only a few people;
- Parametric modeling is an approach that should be used for products with many variants and variations, as is the case with the considered excavator buckets.

6. Application of the Parametric Modeling Approach

The presented parametric modeling approach is applied in the development of various products, which demonstrates its functionality and versatility. The following figures will present examples of products developed using the parametric modeling method through configurations.

For each of the examples, a single 3D model has been developed. In the assembly of the product, the parts and sub-assemblies are parametrically modeled with configurations.

Figures 27 and 28 show cleaning buckets [1,2]. They are parametrically modeled with configurations. The buckets vary in length—from 1600 mm to 2400 mm. Each configuration represents a different bucket length.

Figure 29 shows trapezoidal buckets [1,2]. They are parametrically modeled with configurations. The buckets vary in length—from 2500 mm to 5000 mm. Each configuration represents a different bucket length.

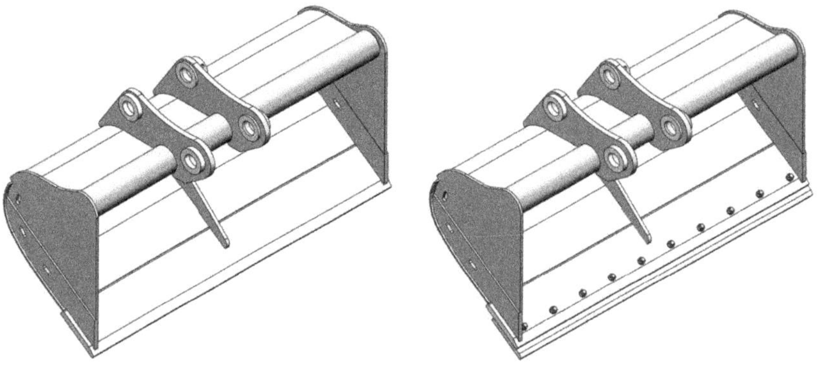

(**a**) (**b**)

Figure 27. Parametric models of cleaning buckets for straight terrain, class D: (**a**) 2000 mm; (**b**) 2000 mm, with additional blade.

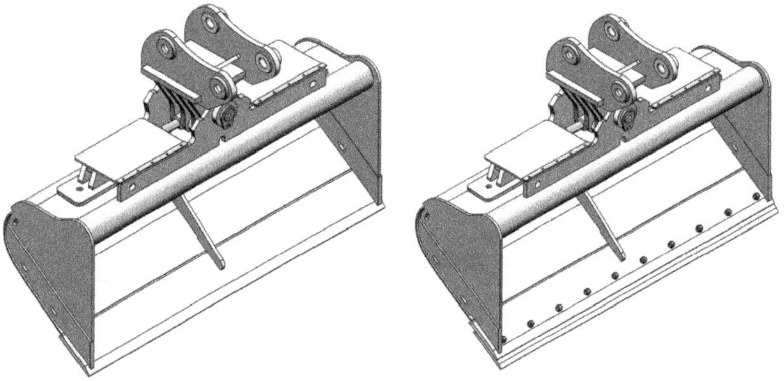

(**a**) (**b**)

Figure 28. Parametric models of cleaning buckets for sloping terrain, class D: (**a**) 2000 mm; (**b**) 2000 mm, with additional blade.

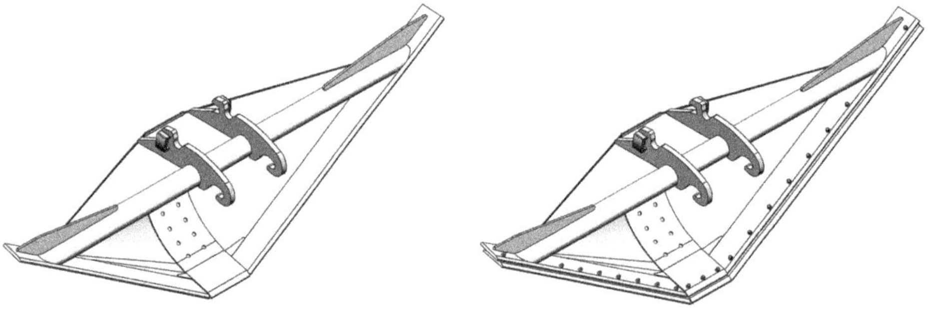

(**a**) (**b**)

Figure 29. Trapezoidal buckets: (**a**) 5000 mm; (**b**) 5000 mm, with additional blades.

Figure 30 shows the scarifier tooth [1,2]. It is parametrically modeled with configurations. The products differ in length. Each configuration represents a different length of the scarifier tooth.

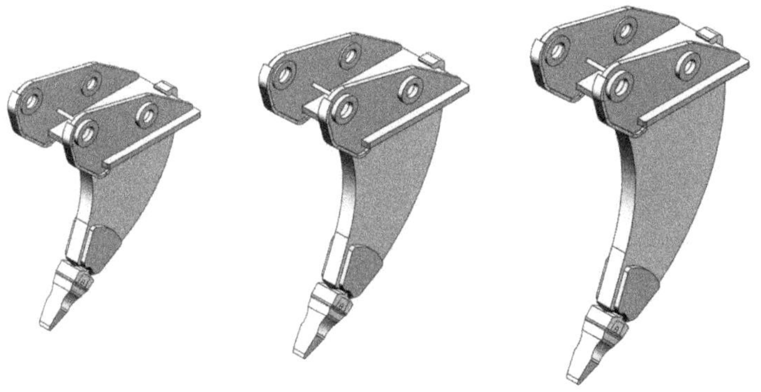

(**a**) (**b**) (**c**)

Figure 30. Scarifier tooth (ripper): (**a**) 950 mm; (**b**) 1150 mm; (**c**) 1350 mm.

The following figures show grapple buckets, developed again using the proposed method for parametric modeling by configurations [26–28]. The grapple bucket consists of left and right buckets, which are manufactured as a set (Figure 31).

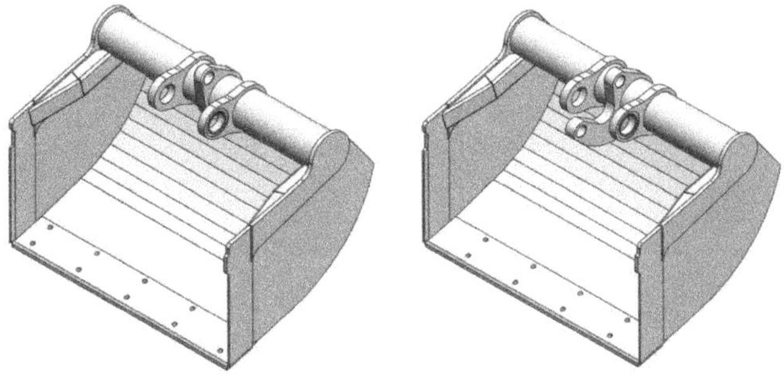

(**a**) (**b**)

Figure 31. Components of the grapple bucket, 1000 mm: (**a**) left bucket; (**b**) right bucket.

Figure 32 shows the 3D model of the bucket, which is parametrically modeled. The buckets differ in length, respectively—600 mm, 800 mm, and 1000 mm.

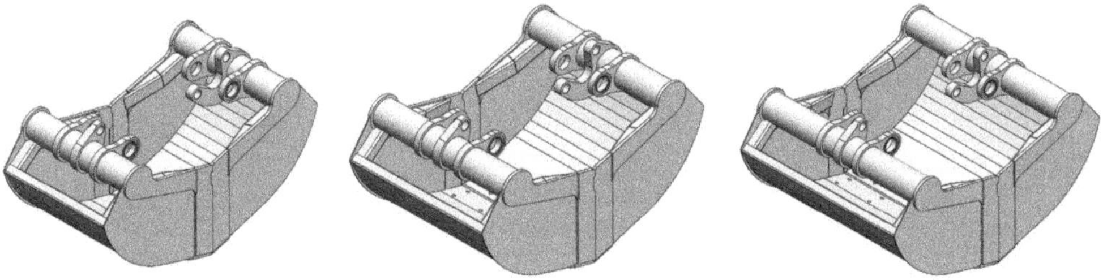

(**a**) (**b**) (**c**)

Figure 32. Parametric grapple bucket models: (**a**) 600 mm; (**b**) 800 mm; (**c**) 1000 mm.

The grapple buckets have a complex shape and construction. For the manufacture of the left and right grapple buckets in a set, specially designed fixtures are used. The fixture is used to base the parts and sub-assemblies of the buckets in a specific position with precisely defined distances between them. Figure 33 shows the designed fixture for assembling the grapple buckets, which is also parametrically modeled.

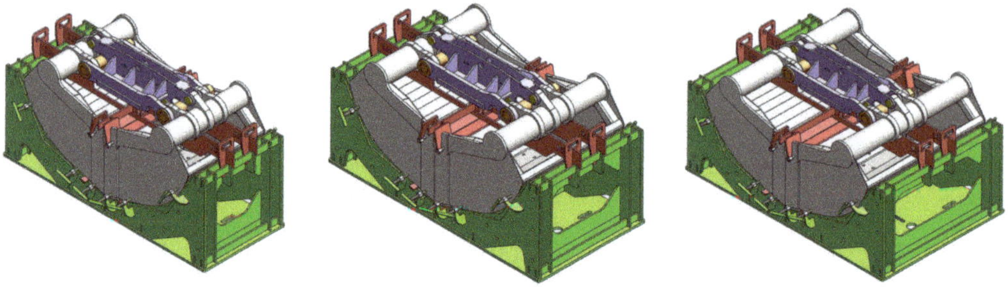

(**a**) (**b**) (**c**)

Figure 33. Parametric models of the grapple bucket manufacturing fixture: (**a**) 600 mm; (**b**) 800 mm; (**c**) 1000 mm.

Figures 34–38 show other examples designed using the parametric modeling approach.

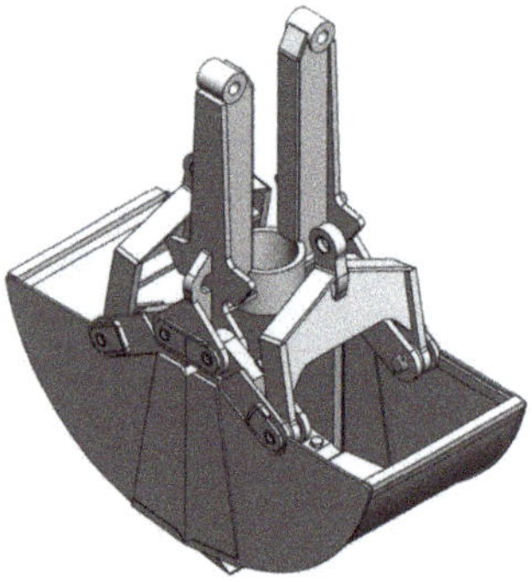

Figure 34. Parametric model of a grapple for bulk materials, 400 mm.

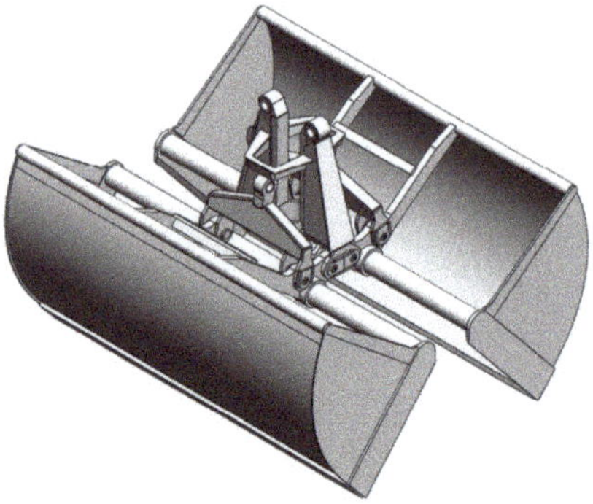

Figure 35. Parametric model of a grapple for bulk materials, 2600 mm.

The proposed method for parametric modeling through configurations has been successfully validated in real engineering projects, not only in the design of excavator buckets but also in the development of other complex products, such as excavator and crane grapples, crane booms, and hydraulic demolition hammers. All these applications are related to the construction and heavy engineering industries, where parametric modeling provides rapid adaptation of structural elements to different sizes, loads, and installation requirements. In the case of grapples, parametric modeling allows for rapid generation of variants with different sizes of jaws (buckets) and drive mechanisms. In the case of crane booms, parametric modeling provides various boom sizes in the set for the crane. In the case of hydraulic hammers, optimization through configurations ensures compatibility with different excavator models and working loads.

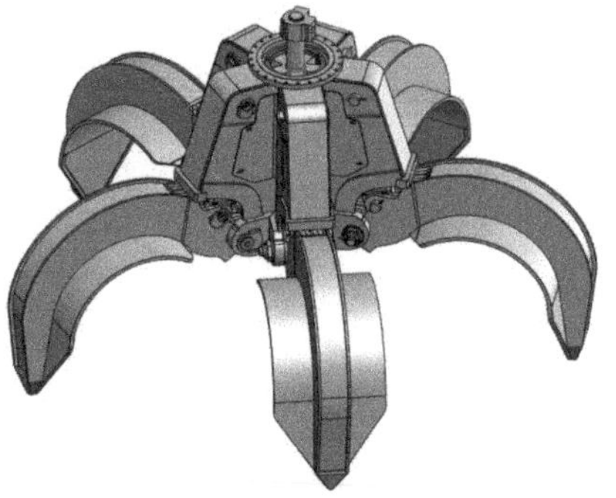

Figure 36. Parametric model of a sorting grapple.

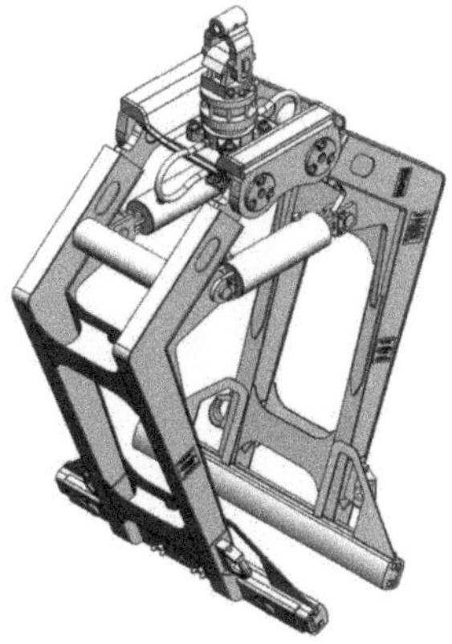

Figure 37. Parametric model of a pallet grabber.

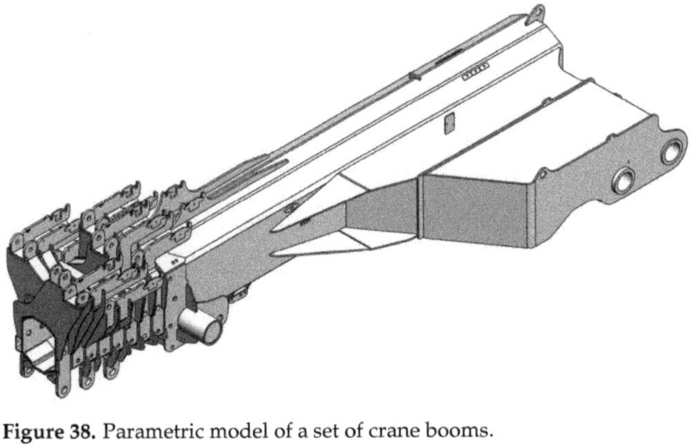

Figure 38. Parametric model of a set of crane booms.

All the products presented in the publication have a complex shape and structure. Specialized fixtures are used to manufacture them. The parametric modeling approach can be used to develop not only products but also specialized fixtures. According to the methodology, only one 3D model of the fixture is developed, which is parametrically modeled using configurations. In the same way, fixtures for assembling, welding, mechanical processing, functional testing, or product control are designed and modeled.

Figures 39–41 show examples of metalworking tools developed using the parametric modeling approach through configurations.

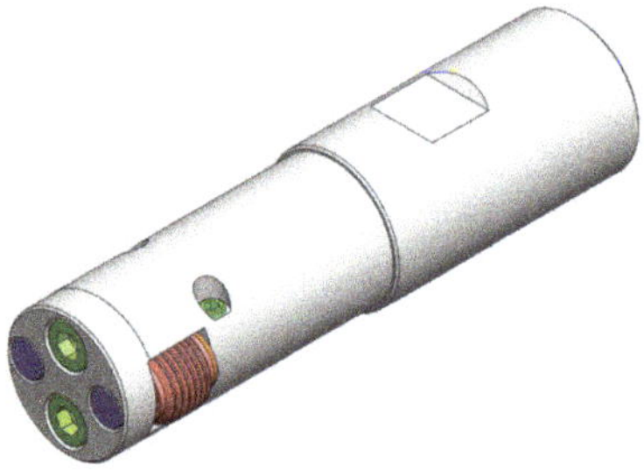

Figure 39. Parametric model of a thread rolling head for internal threads for M30.

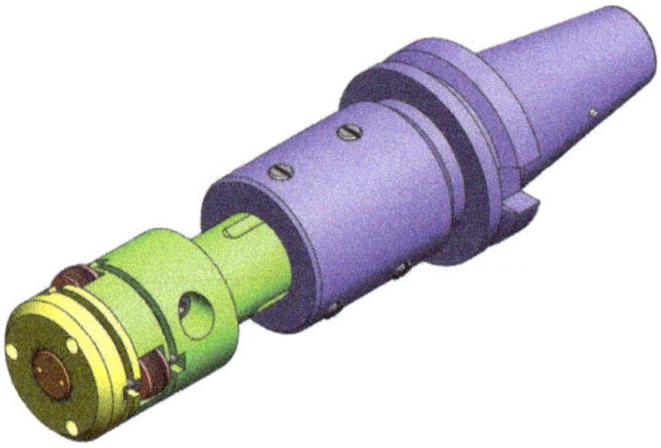

Figure 40. Parametric model of an adjustable thread rolling head.

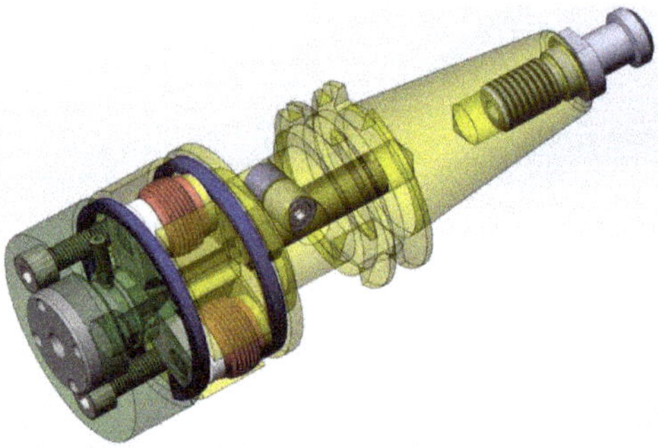

Figure 41. Parametric model of a thread rolling head with automatic retraction of the rolling rolls.

Another place where this methodology finds direct application is the design of metalworking tools. Tools such as thread cutting heads and thread rolling heads have been modeled. The tools consist of a cutting part and shank. According to the specific machine, the shank of the tool is designed through parametric modeling, which ensures its reliable and effective fastening. In this type of product, the main challenge is to provide a standardized but, at the same time, flexible connection between the cutting part of the tool and its shank. The shank is attached to the machine through different holders, each of them has its own standard sizes and design features. Through parametric modeling, configurations can be created that allow the tool to be quickly adapted to different machines, without the need to design the model again. This significantly facilitates the design process and allows the implementation of one tool in different production environments.

The practical implementation of the method in co-operation with industrial partners demonstrates its applicability in various manufacturing sectors and confirms its potential for optimization of engineering processes. In addition, integration with PDM and ERP systems allows effective management of project data and version traceability, which facilitates co-ordination between the different stages of design and production. This not only improves the accuracy of engineering documentation but also reduces the time to implement new products in series production.

7. Discussion

This article presents a methodology for automated design and parametric modeling of excavator buckets, implemented with CAD system SolidWorks 2022 The methodology covers all stages of product design—modeling of 3D models of parts, creating assemblies, and generating design documentation. The article highlights the advantages of this type of parametric design as well as improving work efficiency.

All modern CAD systems offer mechanisms for parametric modeling and configuration, although they have different names and different approaches to implementation. For example, in SolidWorks 2022, this is called "Configurations" [3]; in PTC Creo 11, it is "Family Tables" [29]; in CATIA V6 R2024, it is "Design Tables" [30]; in Autodesk Inventor 2024, it uses "iParts/iAssemblies", automated by iLogic [31]; and Siemens NX 2406 has "Part Families" and "Product Configurations", combined with User-Defined Attributes (UDA) [32]. The methodology considered in this study is not specific to only one CAD system but is applicable to different CAD software. This makes it possible to adapt our approach to different industrial environments, without being limited to the use of only one specific software. Future research could conduct a more in-depth analysis of the performance of these different CAD platforms in industrial applications.

One of the main outcomes from the application of this type of parametric modeling is the significant reduction in the volume of design documentation, as well as the reduction in the complexity of the work performed. Traditional design requires a manual creation of numerous design variations, while the automated generation of different bucket configurations requires minimal human interaction.

The comparison between traditional and parametric modeling through configurations clearly demonstrates a significant advantage in productivity when using configurations. The use of parametric modeling and configurations leads to a significant reduction in the number of necessary 3D models and drawings, which results in 50% time savings in modeling through configurations. Automating the creation of documentation leads to a significant reduction in errors and improves the accuracy of the design. This is especially true for companies where rapid design preparation is required to meet the specific customer requirements.

The proposed methodology has been tested with professional engineers working on real industrial projects commissioned by external companies. They use parametric modeling with configurations within standard industrial processes, which ensures that the measured time and efficiency reflect real operating conditions. Although this study focused on a specific industry area, further testing in different sectors would be beneficial. Future research could extend the evaluation by collecting feedback on the applicability of the method from engineers working in different manufacturing environments, which would allow a wider analysis of its adaptability and effectiveness.

The proposed approach to modeling through configurations has been successfully applied by the R&D Centre at the University of Ruse (Bulgaria) on projects for several different clients. The goal is to increase the efficiency of engineering processes and reduce design time. Engineering departments usually consist of several people and the time required to design and model a specific product should be minimal. Design work is a crucial stage in the product manufacturing process. In real production, the time from receipt of an order for a specific product to its actual production is very short. In this case, the engineering tasks involve more than just developing 3D models and design documentation. It is necessary to know the entire technological process for the product production. The development of 3D models must be consistent with the specific technology for manufacturing the product. The examples of the excavator buckets and all products described in the publication have been manufactured and are actively used in practice. The approach discussed is only part of the entire process of developing a specific product, focusing only on the final 3D modeling and preparation of documentation.

Integrating CAD models with Product Data Management (PDM) and Enterprise Resource Planning (ERP) systems improves the options for data management. By structuring product data within a centralized system, the revision and control processes can be managed effectively. This integration not only optimizes the collaboration between Engineering and Production departments but also improves traceability and adherence to industry standards.

The results of this study are consistent with the existing scientific literature on parametric modeling and its applications in engineering design. Previous studies have highlighted the benefits of automation created by CAD systems to reduce development time cycles and improve product quality. The present methodology builds on these principles, demonstrating their practical application in the design of excavator buckets.

However, although the proposed methodology significantly increases design efficiency, there are also certain limitations. The initial setup and configuration of parametric models require specialized knowledge and thorough understanding of the functionalities of the CAD software. In addition, the effectiveness of the automated design process depends on the stability of the predefined parameters and constraints. It might be exciting to see if we can use AI to optimize the parameterization even more and obtain some better design solutions.

While AI-based modeling has significant potential in engineering design, its direct integration into regulated industrial environments poses a number of challenges, especially related to compliance with regulatory requirements such as the Machinery Directive 2006/42/EC. Current AI solutions are oriented towards optimization according to various criteria such as efficiency. However, we have not found publications that AI solutions have mechanisms for automated compliance with industry standards, traceability of engineering changes, and certification processes. Furthermore, automatically generated AI designs may not pass the necessary regulatory validation for industrial use. These factors make the direct integration of AI into parametric engineering workflows challenging and require additional research. Of particular interest would be the area of AI-based compliance checking and automated regulatory validation.

Any modern publication related to CAD modeling should also address the role of artificial intelligence. Despite the potential of generative AI models in engineering design, their direct integration into the product design process has some significant limitations and challenges. This is especially valid if the designed products are aimed at the EU and other related markets.

Main challenges with using generative AI models in product design:

- Generative AI models create shapes and designs based on optimization algorithms but do not guarantee compliance with industry standards (e.g., ISO, ASME, or the Machinery Directive 2006/42/EC). This creates risks of an invalid product that does not meet the necessary safety, durability, or manufacturability requirements.
- Generative AI models, especially those based on neural networks, often function as a "black box"—it is difficult to understand why the system has made a certain decision. This leads to problems with the traceability of engineering decisions (a key requirement in the Machinery Directive and quality management standards), especially in highly regulated industries such as automotive, aviation, construction tools and equipment, or medical equipment. It is easy to reach a case of a product failure, where it will be difficult to trace which parameter or which logic in the AI led to the defective design.
- CAD systems are designed to work with traditional geometric and parametric models, not with the organic shapes generated by AI. The present AI can create 3D models (STL, STEP, and IGES) that cannot be directly opened in SolidWorks or other 3D CAD systems without additional manual processing [33].
- Generative AI has no engineering intuition or experience, which are often key to making decisions in complex situations. An engineer can predict future work scenarios that AI cannot take into account if they are not included in its training data.

When the safety of machines/devices/equipment directly depends on them being designed using a Large Language Model (LLM), then there will be regulatory implications under the Machinery Directive and its new extended cyber requirements. Here are some scenarios:

- If an error made in the LLM design leads to potential safety risks (e.g., incorrect tolerances, noncompliance with safety standards, unsuitable materials, or incorrect calculations), then control and compliance are required.
- The Machine Directive requires traceability of software changes to designs, and LLM is dynamic and generates code/designs that may be difficult to verify. Validation and certification of the results of each design stage may be required to ensure that LLM does not create risks.
- If the LLM has an impact on the safety of machinery, it is very likely that additional certification or additional inspections will be required (e.g., according to ISO 12100 [7] and IEC 61508 [34]).
- If LLM generates safety code or CAD models that are critical to the safety of the products, validation and testing will need to be provided.

In summary, when LLM is a critical part of the design of machinery, it will require severe additional verifications and compliance with the Machinery Directive. Depending on the use, it may also fall under the new cybersecurity requirements.

Where does AI have potential?

Despite limitations, generative AI technologies have potential as follows:

- For conceptual design—to quickly generate innovative ideas that can be further developed by engineers.
- When optimizing specific parameters—for example, when reducing the weight of parts or minimizing material costs.
- For engineering analysis—discovering correlation dependencies between parameters in complex structures.

In conclusion, this study demonstrates that effective parametric modeling optimizes engineering processes. It reduces design time, minimizes errors, and improves documentation management, supporting the development of digital engineering methodologies. Although the development does not address this issue, the authors believe that the methodology can be applied in other areas of machine design, supporting efficiency and innovation in industrial design.

8. Conclusions

An approach for designing specialized products using a CAD system is presented. The implementation of parametric modeling through the SolidWorks CAD system allows significant automation in the design of excavator buckets. The use of configurations facilitates the creation of various product variants, reducing the time and volume of manually created documentation.

The presented approach is applicable to other similar products, which emphasizes its universal nature. In this way, similar problems related to the development and management of documents can be solved.

The presented bucket design approach provides a possibility for creating a bucket with a new size or a custom bucket design from an existing class within minutes.

Using the parametric modeling approach with configurations, specialized fixtures for manufacturing can be developed. These include fixtures for assembling, welding, mechanical processing, functional testing, product control, etc.

Parametric modeling through configurations is used in various areas of industry where machines, equipment, and structures are designed. The presented methodology is used to model products with multiple variations, sizes, and versions.

The use of attributes (user and systems ones) in the 3D models of parts and assemblies enables the automated preparation of design documentation (drawings of parts and assemblies and specifications).

The proposed approach of modeling through configurations improves the efficiency of engineering work by reducing design time, the volume of design documentation, and errors during product design.

The presented integration between the 3D models and PDM system allows effective management of a large volume of design documentation. Automatic synchronization of attributes in models and the system leads to better traceability, variation, and accessibility of data for all departments in the company, including departments such as Supplies and Production. The files can also be used by other company departments through the ERP system. This is an innovative approach to the digitalization of engineering processes.

The use of specialized CAM software to create machining programs significantly increases the accuracy and efficiency in bucket production. The inclusion of 3D models for direct preparation of production documentation demonstrates the modernization of traditional production methods, while facilitating the implementation of technological improvements.

Although the present study demonstrates the advantages of parametric modeling in the automation of the design process for excavator buckets and other similar products, the potential of this approach can be extended in several directions.

One possible future direction is the integration of the proposed methodology with the concept of Digital Twins. The parametric models that have been developed can be used as a basis for building virtual representations of the products, which can be updated in real time with operational data. This would allow tracking of wear and load on structures, as well as optimization of future designs based on real operating conditions.

Another important aspect is the potential use of artificial intelligence for automatic generation and optimization of parametric models. AI-based algorithms could suggest optimal design solutions based on analysis of previous designs and operational data. Future research could explore the interaction between parametric modeling and AI to improve the flexibility and efficiency of the design process. Changes in regulations in major global markets regarding AI will lead to additional challenges for the use of AI in the design of industrial products.

The main challenge in using generative AI models in product design for highly regulated industries (automotive, aerospace, construction tools and equipment, medical devices, etc.) is the lack of ability for control, traceability, and compliance with market regulations and industry standards. While these systems can be a powerful tool for the conceptual phase of designing or optimizing specific aspects, they are not yet able to replace human engineering judgment and strictly regulated engineering processes.

Author Contributions: Conceptualization, N.S. and A.I.; methodology, N.S. and A.I.; software, N.S.; validation, N.S. and A.I.; formal analysis, N.S.; investigation, N.S.; resources, N.S.; data curation, N.S. and A.I.; writing—original draft preparation, N.S.; writing—review and editing, N.S. and A.I.; visualization, N.S.; supervision, N.S.; project administration, N.S.; funding acquisition, N.S. and A.I. All authors have read and agreed to the published version of the manuscript.

Funding: This study is financed by the European Union—NextGenerationEU, through the National Recovery and Resilience Plan of the Republic of Bulgaria, project No. BG-RRP-2.013-0001.

Institutional Review Board Statement: Not applicable.

Informed Consent Statement: Informed consent was obtained from all subjects involved in the study.

Data Availability Statement: The raw data supporting the conclusions of this article will be made available by the authors on request.

Conflicts of Interest: The authors declare no conflicts of interest.

References

1. Buckets Line. 2025. Available online: https://mantovanibenne.com/en/attachments/buckets/ (accessed on 12 March 2025).
2. Buckets Line. 2025. Available online: https://www.mbibulgaria.bg/ (accessed on 12 March 2025).
3. SolidWorks, 3D Design Software. SolidWorks CAM, CAM Software. SolidWorks PDM, PDM Software. 2025. Available online: https://www.solidworks.com/ (accessed on 12 March 2025).
4. RDS ERP, ERP Software. 2025. Available online: https://www.smeup.com (accessed on 12 March 2025).
5. Stankov, N.; Ivanov, A. *Design Documentation Management with SolidWorks Enterprise PDM. Collection of Scientific Works*; University of Ruse: Ruse, Bulgaria, 2015; Volume 54, Series 2, pp. 158–165, ISSN 1311-3321.
6. Truhcheva, D.; Stankov, N.; Ivanov, A. *Management and Organization of Design Documentation with SolidWorks and SolidWorks Enterprise PDM. Collection of Reports of a Student Scientific Session–SSS'15*; University of Ruse: Ruse, Bulgaria, 2015; pp. 20–30. ISSN 1311-3321.
7. *ISO 12100:2010*; Safety of Machinery—General Principles for Design—Risk Assessment and Risk Reduction. International Organization for Standardization: Geneva, Switzerland, 2010.
8. Autodesk Inventor. 3D Design Software. 2025. Available online: https://www.autodesk.com/ (accessed on 12 March 2025).
9. Solid Edge. 3D Design Software. 2025. Available online: https://solidedge.siemens.com/en/ (accessed on 12 March 2025).
10. Creo. 3D Design Software. 2025. Available online: https://www.ptc.com/en (accessed on 12 March 2025).
11. CATIA. 3D Design Software. 2025. Available online: https://www.3ds.com/products/catia (accessed on 12 March 2025).
12. Altair Inspire. 3D Design Software. 2025. Available online: https://altair.com/inspire (accessed on 12 March 2025).
13. Stankov, N. *Design of Excavator Buckets with the SolidWorks CAD System. Collection of Reports of a Student Scientific Session–SSS'10*; University of Ruse: Ruse, Bulgaria, 2010; pp. 16–21, ISSN 1311-3321.
14. Pecingina, O. Modelling bucket excavation by finite element. *IOP Conf. Series: Mater. Sci. Eng.* **2015**, *95*, 012046. [CrossRef]
15. Vîlceanu, F.; Iancu, C. Bucket wheel rehabilitation of ERC 1400-30/7 high-capacity excavators from lignite quarries. *IOP Conf. Series Mater. Sci. Eng.* **2016**, *161*, 012081. [CrossRef]
16. Bošnjak, S.; Arsić, M.; Gnjatović, N.; Milenović, I.; Arsić, D. Failure of the bucket wheel excavator buckets. *Eng. Fail. Anal.* **2018**, *84*, 247–261. [CrossRef]
17. Ren, Z.; Sun, H.; Liang, Y.; He, Y.; Feng, M. Strength analysis of excavator bucket based on normal digging trajectory and limiting digging force. *J. Vibroengineering* **2020**, *23*, 217–226. [CrossRef]

18. Suryo, S.H.; Sastra, R.S.; Muchammad, H. Optimization of bucket tooth excavator design using topology optimization and finite element method. *J. Physics Conf. Ser.* **2021**, *1858*, 012081. [CrossRef]
19. Hepa, M. Design and Optimization of Lightweight and HSD Excavator Bucket with Uncertain Load. *Int. J. Curr. Eng. Technol.* **2021**, *11*, 222–226. [CrossRef]
20. Sun, Y.; Wang, Y.; Wang, L.; Li, C.; Tang, L.; Wang, D.; Ma, R.; Xue, Z.; Wei, X.; Cui, M.; et al. Digging Performance and Stress Characteristic of the Excavator Bucket. *Appl. Sci.* **2023**, *13*, 11507. [CrossRef]
21. Wang, Y.; Wang, L.; Li, C.; Xue, Z.; Sun, Y.; Ma, R.; Wang, D.; Cui, M.; Wei, X.; Tang, L.; et al. Optimization of Excavator Bucket Structure by a Coupled Simulation Method. *Appl. Sci.* **2023**, *13*, 11336. [CrossRef]
22. Liu, G.; Han, X.; Wang, Z.; Wang, K.; Zhang, Z.; Duan, Z. Viscosity Reduction and Drag Reduction Performance Analysis of Bionic Excavator Buckets Based on Discrete Element Method. *Biomimetics* **2024**, *9*, 686. [CrossRef] [PubMed]
23. Abebe, M.; Cho, Y.; Han, S.C.; Koo, B. Mitigating Measurement Inaccuracies in Digital Twins of Construction Machinery through Multi-Objective Optimization. *Sensors* **2024**, *24*, 3347. [CrossRef] [PubMed]
24. Liu, C.; Li, Y.; Jiang, X.; Shao, W. Five-axis flank milling tool path generation with curvature continuity and smooth cutting force for pockets. *Chin. J. Aeronaut.* **2020**, *33*, 730–739. [CrossRef]
25. Wang, Y.; Wang, D.; Zhang, S.; Tang, Z.; Wang, L.; Liu, Y. Design and development of a five-axis machine tool with high accuracy, stiffness and efficiency for aero-engine casing manufacturing. *Chin. J. Aeronaut.* **2022**, *35*, 485–496. [CrossRef]
26. High Volume Clamshell Buckets. 2025. Available online: https://www.kinshofer.com/en/crane/general-construction/high-volume-clamshell-buckets (accessed on 12 March 2025).
27. SL Industries. 2025. Available online: https://sl-industries.com/ (accessed on 12 March 2025).
28. Bullitt Engineering. 2025. Available online: https://bullitt-engineering.com/ (accessed on 12 March 2025).
29. Family Tables. 2025. Available online: https://support.ptc.com/help/creo/creo_pma/r11.0/usascii/index.html#page/fundamentals/fundamentals/About_Family_Tables_1.html (accessed on 12 March 2025).
30. Managing CATIA V5 Design Tables. 2025. Available online: https://support.ptc.com/help/windchill/plus/r12.0.2.0/en/index.html#page/Windchill_Help_Center/WGMCATIAV5AdvTechFamTable.html (accessed on 12 March 2025).
31. iPart and iAssembly Functions Reference (iLogic). 2025. Available online: https://help.autodesk.com/view/INVNTOR/2025/ENU/?guid=GUID-4BA100AA-B55A-4A08-AD8F-79AA27771C7E (accessed on 12 March 2025).
32. Managing NX Part Families. 2025. Available online: https://support.ptc.com/help/windchill/plus/r12.1.2.0/en/index.html#page/Windchill_Help_Center/WWGMforNX/WWGMNXAdvTechNXManagePartFams.html (accessed on 12 March 2025).
33. Kozov, V.; Ivanov, A.; Ivanova, G. Research Result Differences Between Manual Research and Using AI LLMs for Analyzing Popular 3D File Formats. *TEM J.* **2024**, *13*, 2637–2658. [CrossRef]
34. Iec 61508; Safety and Functional Safety. Available online: https://www.iec.ch/functional-safety (accessed on 12 March 2025).

applied sciences

MDPI

Article

Analytical and Graphical Profiling of Thread-Milling Cutters for Forming Internal Threads

Nikolay Stankov * and Aleksandar Ivanov *

Department of Machine Tools and Manufacturing, Faculty of Mechanical and Manufacturing Engineering, University of Ruse, 8 Studentska Street, 7017 Ruse, Bulgaria
* Correspondence: nstankov@uni-ruse.bg (N.S.); akivanov@uni-ruse.bg (A.I.)

Abstract: Accurate profiling of thread-milling cutters is one of the main prerequisites of high-precision internal thread production. Despite the accuracy of analytical methods, their complexity in most instances makes practical application an issue. The current research addresses a graphical profiling method, the Tangent Circles Method, introduced by the authors for the facilitation of thread-milling tool geometric design. Several thread-milling cutters were designed employing this method, and their working surface profiles were compared with those obtained analytically. The comparison showed a good match, with profile deviations within acceptable manufacturing limits. CAD-based simulations also demonstrated that rake angle and relief height parameters significantly influence the resulting tool geometry. The results validate the Tangent Circles Method as a mathematically sound and industrially viable tool-profiling method for industry use. Its easy application and precision make it a reliable choice over analytical methods in internal thread-milling cutters' industrial design.

Keywords: internal threads; thread-milling cutters; tool surface; analytical profiling; graphical profiling; CAD system

1. Introduction

Thread milling is among the common processes of internal thread manufacturing in industrial applications due to its flexibility, surface finish parameters and suitability in CNC machine centers. Among the various categories of tools that are available, some special thread-milling cutters are more applicable for cutting internal threads in complex or non-standard parts.

Despite their advantages, the biggest problem in their conception is precise profiling of the cutter's working areas. The cutter must exactly model the thread shape without inducing undercutting of flanks. Analytical profiling methods, although precise, are prone to be heavily mathematical and computation-intensive. Graphical methods, however, are easier to carry out but are hardly ever verified to design high-precision tools.

The graphical profiling technique developed by the authors, known as the Tangent Circles Method, is applied in this work to define the geometry of cutting tools with complex profiles. This graphical method was originally developed by the authors as a practical approach to tool profiling and is presented here along with its validation against analytical results. To the best of the authors' knowledge, this is the first time the Tangent Circles Method is formally described and tested in the context of internal thread-milling tool design.

Several thread-milling cutters are designed, and the tool profiles obtained using the graphical method are directly compared with those generated analytically. This comparison allows us to assess the mathematical accuracy of the graphical approach. Additionally, CAD-based simulations are used to evaluate the influence of tool design parameters such as rake angle and relief height on the final shape of the tool profile. The aim of this study is to validate the Tangent Circles Method as a reliable and precise alternative to analytical profiling for internal thread-milling cutters and to examine how design variables affect tool geometry through simulation.

2. Literature Review

2.1. Thread-Forming Methods

Thread forming is a complex and laborious technological process. Nevertheless, the formation of threads and other helical surfaces is one of the most common technological tasks in mechanical engineering. Threads are widely used in various designs of machines, mechanisms, devices and many differently shaped parts and fasteners.

The wide variety of machine parts with different shapes, sizes and accuracies of helical surfaces, as well as the variety of materials from which they are made, also determines the variety of methods for forming threads, which are reduced to the following three groups:

- Methods for thread forming by chip removal;
- Methods for forming threads by plastic deformation of the processed material in a non-heated state;
- Methods for forming threads during the primary forming of parts.

The first two methods are the main and most common ones. Their choice, in addition to the accuracy, productivity and efficiency provided, mainly depends on the values of the mechanical characteristics of the material being processed—Brinell hardness (HB), tensile strength (R_m) and relative elongation (A%).

Cutting processes provide higher quality characteristics of the machined surfaces, high accuracy, low roughness and higher strength, yield and fatigue limits of the machined parts.

The third group of methods forms threaded surfaces or parts of threaded surfaces in hard alloy and plastic parts. The methods used are powder metallurgy, pressing or pressure casting, etc.

Figure 1 shows a flowchart with the main characteristics determining the applicability of the methods of forming threads [1,2].

2.2. Forming of Internal Threads by Milling

Forming internal threads by cutting is the most common method. Internal threads can be formed with thread-turning tools, taps, disk thread-milling cutters, thread-milling cutters and thread-cutting heads [1,2]. In practice, forming internal threads is mainly realized with thread-milling cutters.

Short internal threads with fine and normal pitches are made with thread-milling cutters. The method of cutting threads with thread-milling cutters is performed on special thread-milling machines or CNC machines under conditions of medium and mass production. These tools in certain cases successfully replace thread-cutting taps, especially when cutting threads in body parts, due to the fact that when the tool is destroyed, the geometric characteristics of the machined threaded surfaces remain intact. The disadvantages of this type of tool include the deviations in the profile of the formed thread due to the thread-forming scheme and the design features of the tools. These deviations are reduced by implementing a new working profile of the cutters determined by profiling.

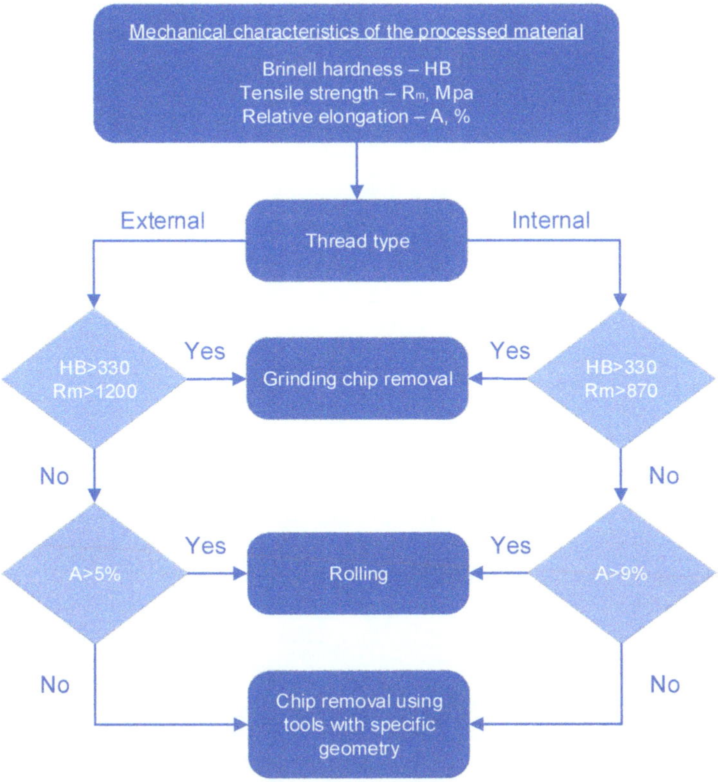

Figure 1. Flowchart for determining the applicability of the most common groups of thread forming methods.

During operation, the tool axis is parallel to the workpiece axis. The working movements performed by the thread-milling cutter are shown in Figures 2 and 3. Milling of internal threads is realized by the main rotary movement 1, radial feed movement 2, circular feed movement 3 and longitudinal feed movement 4. The thread-milling cutter cuts radially into the workpiece with movement 2 until the complete profile of the thread is formed. The thread is formed by up-milling while performing a circular feed movement 3 and axial movement of the tool with a movement 4 equal to the thread pitch P.

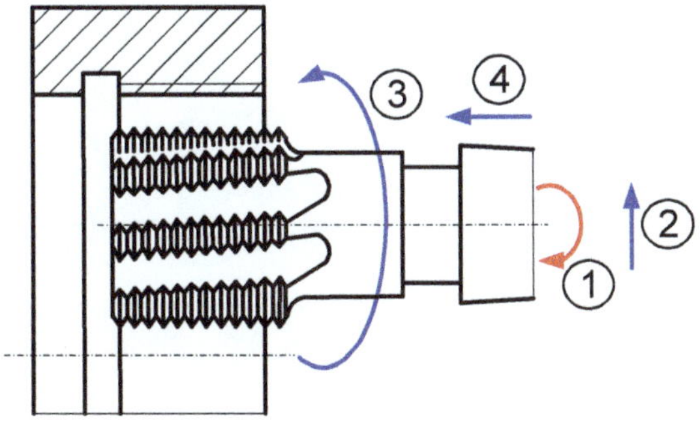

Figure 2. Working movements when milling an internal thread with a thread-milling cutter in a cross-section of the workpiece.

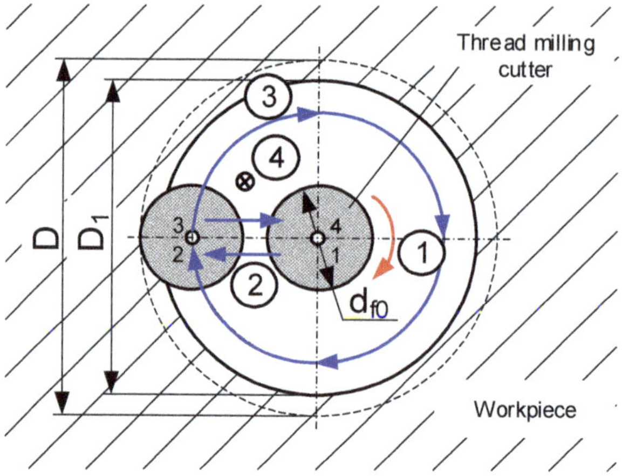

Figure 3. Working movements when milling an internal thread in an axial section of the workpiece.

The forming of internal threads by milling is performed on CNC machines, and specifically on machining centers, because of their capabilities to provide the complex combinations of main and feed-forming movements. CNC systems contain a variety of automated cycles for thread processing. Through them, high accuracy and efficiency of processing is achieved, significantly reducing machine times [3–5].

Despite the many sources of information on the process of milling internal threads with thread-milling cutters, there are many issues that have not been sufficiently addressed. For example, questions about solving the problem of determining the tool profile to compensate for deviations from the theoretical profile due to the parallelism of the axes of the tool and the workpiece, the formation of helical chip flutes or the presence of a rake angle. There are also issues related to technological and design errors when milling internal threads, etc.

Figure 4 shows the methods used for profiling the working part of metal-cutting tools.

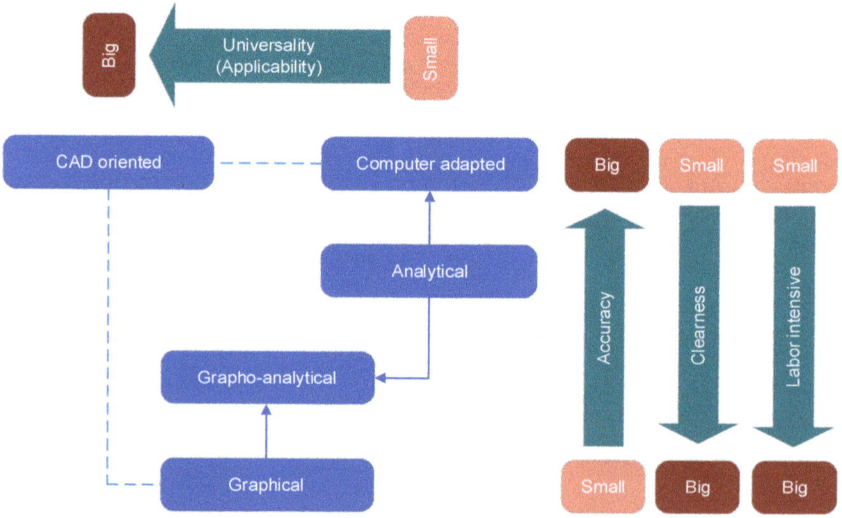

Figure 4. Profiling methods.

The classical methods for profiling cutting tools are the grapho-analytical methods. In this type of design, part of the input data for the analytical part of the profiling is taken from graphical schemes developed on a scale with high magnification. However, their accuracy is relatively low. Grapho-analytical methods are used to verify the results of classical analytical modeling.

With the advent of modern CAD systems for three-dimensional modeling, the need for such scaling disappears because it is possible to work with the real dimensions of the products, and the accuracy with which the graphic schemes are measured, for example in the CAD system SolidWorks, is eight digits after the decimal point. The required accuracy, which is usually used when profiling the cutting tools, is high, at four digits after the decimal point. In this way, the use of modern CAD systems completely replaces the classical methods of grapho-analytical design, and there is no need to check the results analytically.

Profiling is used in the manufacture of various types of metal-cutting tools. Through analytical or graphical profiling, the working parts of the tools are obtained, which in most cases represent surfaces with a complex shape. Such tools are, for example, thread-turning tools [1,2], taps [1,2], dies [1,2], thread-cutting heads [1,2], finger-milling cutters [1,2], end mills [1,2,6–9], thread-milling cutters [1,2,10], disk-shaped thread-milling cutters [1,2,11], ring-shaped tools for vortex thread milling [12], combined tools [13,14], profile turning tools [15–17], gear-shaped cutters [1,2,18], profile milling cutters [19], single-tooth thread-milling cutters [20] etc.

The publication considers the profiling of the working part of thread-milling cutters by an analytical method and by a graphical method using a CAD system. The profiling of tools for forming metric threads is considered. These are the threads that are widely used in practice. If a tool needs to be designed for another type of thread, the proposed method for graphic profiling can be used. The method is applicable for all types of threads that have a straight profile.

3. Analytical Determination of the Tool Surface Profile of Thread-Milling Cutters for Milling Internal Threads

When milling threads with thread-milling cutters, the tools are oriented relative to the machined surface so that the axis of the rotary profile working part of the tool is parallel to the geometric axis of the machined thread. This positioning of the tool relative to the milled surface leads to deviations in the axial profile of the latter. The forming of the specified axial profile of the formed surface without deviations is achieved by implementing special profiling of the tool surface.

To determine the profile of the tool surface on which the cutting edges of the teeth are located, with a known profile of the formed surface, an analytical method was used. Using a graphical method, the same profiling of the tool surface was performed through the SolidWorks 2022 CAD system, and its accuracy and reliability were validated.

3.1. Methodology for Analytical Determination of the Profile of the Tool Surface

Figure 5 shows the profile of an internal thread in cross- and axial section.

A thread whose profile in axial section is linear is defined as an Archimedean helical surface. Any section of it normal to its geometric axis, for example, section A-A in Figure 5, is an Archimedean spiral with an equation in polar coordinates:

$$R_i = a.\varphi_i, \text{mm} \quad (1)$$

where R_i is the radius vector of the Archimedean spiral in mm; a is parameter of the Archimedean spiral in mm/rad; φ_i is the angle of rotation of the spiral in rad.

These cross-sections are identical but are rotated relative to each other by an angle ε_i, depending on the helical parameter of the helical surface (Figure 6):

$$\varepsilon_i = \frac{z_i}{p}, \text{rad} \quad (2)$$

where $p = \frac{n.P}{2.\pi}$ is a helical parameter; P is the thread pitch in mm; n is the number of thread passes; z_i is the displacement of the secant plane along the "z" axis in mm.

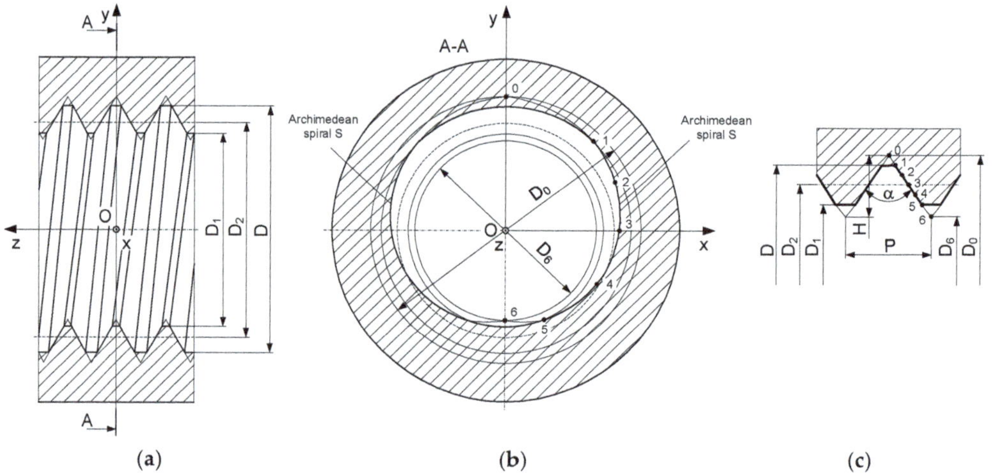

Figure 5. Profile of the formed thread: (**a**) cross-section; (**b**) axial section; (**c**) thread parameters.

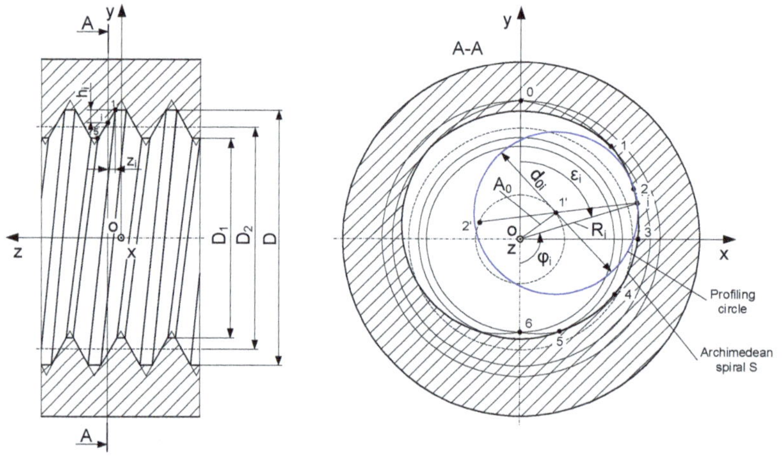

Figure 6. Scheme for determining the profile of the tool surface when milling internal threads.

The secant plane A-A also intersects the rotational tool surface oriented in the "Oxyz" coordinate system. Its section in the same secant plane is a circle with diameter d_{0i}, which is tangent to the Archimedean spiral S at point i, where they have a common normal passing through points i and 1′ (Figure 6). This section is a profiling section of the tool surface. Since the rotary tool surface has a smaller diameter than the inner diameter of the thread, the center distance A_0 is set, which is determined by the following equation:

$$A_0 = R - r_0, \text{mm} \quad (3)$$

where R is the radius of the outer diameter of the thread and r_0 is the radius of the circle of the tool tangent to the outer diameter of the thread in mm.

The circle drawn with radius A_0 is the trajectory of the circular feed motion 3 (Figures 2 and 3). In order to realize the milling, it is necessary that the normal vector at any point of the spiral describing the cross-section of the thread intersects this circle.

Considering that with known pitch P, number of thread passes n, average diameter D_2 and thread profile angle α each helical surface is determined, then Archimedean spirals are also determined, drawing the different sections that the tool touches. Their parameter "a" can be determined by the following equations:

$$R_0 = a.\varphi_0, \text{mm} \quad (4)$$

$$R_6 = a.\varphi_6, \text{mm} \quad (5)$$

$$H = R_0 - R_6 = \frac{P}{2.tg\frac{\alpha}{2}}, \text{mm} \quad (6)$$

where R_0 is the radius vector of point 0 of the Archimedean spiral drawn at angle φ_0 in mm; R_6 is the radius vector of point 6 on the same spiral, but at an angle φ_6 in mm; H is the height of the theoretical thread profile in mm.

Then,

$$H = a.\varphi_0 - a.\varphi_6 = a.(\varphi_0 - \varphi_6), \text{mm} \quad (7)$$

The radius vectors of the spirals increase from R_6 to a value of R_0 for an angle $\varphi_6 - \varphi_0 = \frac{\pi}{n}$. Then, from Equation (7), it follows that $H = \frac{a.\pi}{n}$, and taking into account Equation (6), for the spiral parameter, the following equation is obtained:

$$a = \frac{n.P}{2\pi.tg\frac{\alpha}{2}}, \text{m/rad} \quad (8)$$

For a certain parameter of the spiral "a" and given radius vectors R_i at different points in the range $R_1 \div R_6$, the angle of rotation of the spiral φ_i is determined. For each value of φ_i, the position of the secant plane z_i in the coordinate system "Oxyz" is also determined by the following equation (Figure 7):

$$z_i = \frac{P}{4} - (R_i - R_2).tg\frac{\alpha}{2}, \text{mm} \quad (9)$$

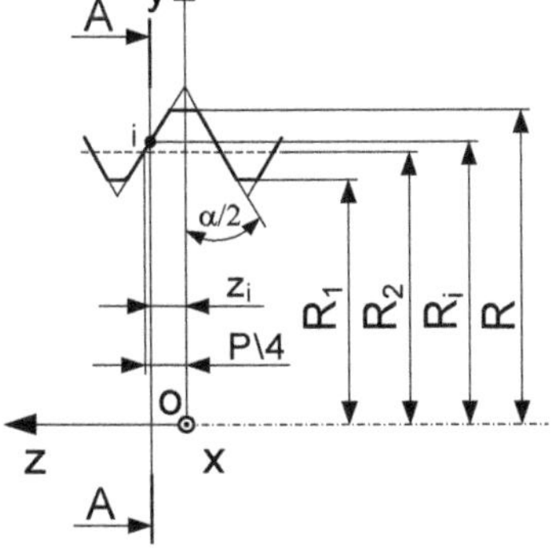

Figure 7. Scheme for determining the z_i coordinate of the tool surface.

To determine the radii of the tangent (profiling) circles at any point "i", it is necessary to know the coordinates of the tool center. To determine them, the equation of the circle with radius A_0 describing the trajectory of the center of the tool surface and the equation of the normal to the Archimedean spiral at the i-th point in the "Oxyz" coordinate system is used.

The equation of the circle is as follows:

$$A_0^2 = x^2 + y^2, \mathrm{mm} \tag{10}$$

The equation of the normal at any contact point "i" is determined by the following equation:

$$(y - y_i) = -\frac{dx_i}{dy_i}.(x - x_i) = -k.(x - x_i), \tag{11}$$

where

$$x_i = R_i.\cos\varphi_i, \tag{12}$$

$$y_i = R_i.\sin\varphi_i, \tag{13}$$

$$k = -\frac{a.\cos\varphi_i - R_i.\sin\varphi_i}{a.\sin\varphi_i + R_i.\cos\varphi_i}. \tag{14}$$

By solving (10) and (11) together and taking into account (12)–(14), after processing, the following quadratic equation is obtained:

$$\left(1 + k^2\right).x_0^2 + 2.k.(y_i - k.x_i).x_0 + k^2.x_i^2 + y_i^2 - A_0^2 - 2.k.x_i.y_i = 0. \tag{15}$$

Taking into account that

$$a_1 = 1 + k^2, \tag{16}$$

$$b_1 = 2.k.(y_i - k.x_i), \tag{17}$$

$$c_1 = k^2.x_i^2 + y_i^2 - A_0^2 - 2.k.x_i.y_i = \left(k.x_i - y_i\right)^2 - A_0^2, \tag{18}$$

Equation (15) becomes

$$a_1.x_0^2 + b_1.x_0 + c = 0. \tag{19}$$

The elements of the equation are as follows:

$$x_{1'} = \frac{-b_1 + \sqrt{b_1^2 - 4.a_1.c_1}}{2.a_1}, \tag{20}$$

$$x_{2'} = \frac{-b_1 - \sqrt{b_1^2 - 4.a_1.c_1}}{2.a_1}. \tag{21}$$

These are the abscissas of the intersection points $1'$ and $2'$ of the normal with the trajectory of the tool center (Figure 6).

Using Equation (11) with known abscissas $x_{1'}$ and $x_{2'}$, the values of the ordinates of the intersection points $y_{1'}$ and $y_{2'}$ are also determined. In order to determine the diameter of the profiling circle d_{0i}, the distances $l_{1'}$ and $l_{2'}$ between these points and point "i" are determined according to the following equations:

$$l_{1'} = \sqrt{(x_{1'} - x_i)^2 + (y_{1'} - y_i)^2}, \tag{22}$$

$$l_{2'} = \sqrt{(x_{2'} - x_i)^2 + (y_{2'} - y_i)^2}. \tag{23}$$

The radius of the profiling circle r_{0i} is assumed to be equal to the smaller value of $l_{1'}$ or $l_{2'}$.

The forming line of the profile of the disk tool is obtained from the series of values of z_{0i} and r_{0i} and drawing the symmetrical profile (Figure 8).

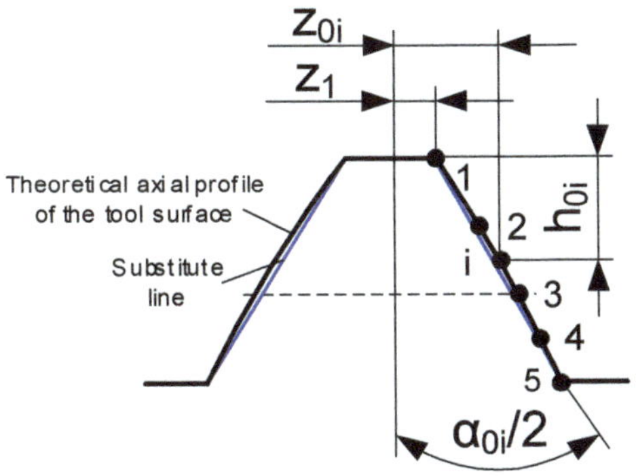

Figure 8. Tool surface profile.

By ignoring the slight curvature of the side forming the profile and replacing it with a straight line that passes through points 1 and 5, the profile angle of the axial profile of the producing tool surface is also determined (Figure 8).

After replacing the resulting profile curve with a straight line passing through points 1 and 5, the difference between the thread profile angle α and the tool surface profile angle α_{0i} (i = 5) is determined. This difference is taken as the necessary correction $\Delta\alpha_0$ of the tool surface profile and is determined by the following equations:

$$z_{0i} = z_i - z_{1,mm} \tag{24}$$

$$h_i = R_{imax} - R_{i,mm} \tag{25}$$

$$h_{0i} = r_{0imax} - r_{0i,mm} \tag{26}$$

$$\alpha_{0i} = 2.\text{arctg}\left(\frac{z_{0i}}{h_{0i}}\right), ^\circ \tag{27}$$

$$\Delta\alpha_0 = \alpha - \alpha_{0i}, ^\circ \tag{28}$$

where z_1 is the applicate of point 1 of the profile, determined by Equation (9) at i = 1 in mm; R_{imax} is the maximum radius of all radius vectors R_i, determined by Equation (1), in mm; r_{0imax} is the maximum radius of all radii of the profiling circle r_{0i}, determined by Equations (22) and (23), in mm.

Using the developed analytical profiling methodology, a Microsoft Excel 2019 program was created to determine the profile of the tool surface. It is assumed that the coordinates of five points of the thread profile are calculated (Figure 8).

Table 1 summarizes the methodology for determining the parameters of the tool surface profile. The parameters and equations used to determine them are shown.

Table 1. Equations for determining the parameters of the tool surface profile.

Parameter	Equation
Tool surface center trajectory radius	$A_0 = R - r_0$, mm
Parameter of the Archimedean spiral	$a = \frac{n.P}{2\pi.\text{tg}\frac{\alpha}{2}}$
Current radius vector of the spiral	$R_i = \frac{D_i}{2}$, mm

Table 1. *Cont.*

Parameter	Equation
Spiral rotation angle	$\varphi_i = \frac{R_i}{a}, \text{rad}$
Abscissa of point i	$x_i = R_i.\cos\varphi_i, \text{mm}$
Ordinate of point i	$y_i = R_i.\sin\varphi_i, \text{mm}$
Angular coefficient of the normal in point i	$k = -\frac{a.\cos\varphi_i - R_i.\sin\varphi_i}{a.\sin\varphi_i + R_i.\cos\varphi_i}$
Parameter a_1	$a_1 = 1 + k^2$
Parameter b_1	$b_1 = 2.k.(y_i - k.x_i)$
Parameter c_1	$c_1 = \left(k.x_i - y_i\right)^2 - A_0^2$
Abscissa of point 1′	$x_{1'} = \frac{-b_1 + \sqrt{b_1^2 - 4.a_1.c_1}}{2.a_1}, \text{mm}$
Abscissa of point 2′	$x_{2'} = \frac{-b_1 - \sqrt{b_1^2 - 4.a_1.c_1}}{2.a_1}, \text{mm}$
Ordinate of point 1′	$y_{1'} = k.(x_{1'} - x_i) + y_i, \text{mm}$
Ordinate of point 2′	$y_{2'} = k.(x_{2'} - x_i) + y_i, \text{mm}$
Distance between point 1′ and point i	$l_{1'} = \sqrt{(x_{1'} - x_i)^2 + (y_{1'} - y_i)^2}, \text{mm}$
Distance between point 2′ and point i	$l_{2'} = \sqrt{(x_{2'} - x_i)^2 + (y_{2'} - y_i)^2}, \text{mm}$
Radius of the profiled circle	$r_{0i} = \min(l_{1'}, l_{2'}), \text{mm}$
Applicate of point i	$z_i = \frac{P}{4} - (R_i - R_2).\text{tg}\frac{\alpha}{2}, \text{mm}$
Distance between point 1′ and point i along the "y" axis of the profiled circle	$h_{0i} = r_{0imax} - r_{0i,mm}$
Distance between point 1′ and point i along the "z" axis of the profiled circle	$z_{0i} = z_i - z_{1,mm}$
Distance between point 1′ to point i along the "y" axis of the thread	$h_i = R_{imax} - R_{i,mm}$
Difference in profile heights	$\Delta h_i = h_{0i} - h_{i,mm}$
Tool profile angle	$\alpha_{0i} = 2.\text{arctg}\left(\frac{z_{0i}}{h_{0i}}\right), ^\circ$
Deviation in profile angle	$\Delta\alpha_0 = \alpha - \alpha_{0i}, ^\circ$

3.2. Application of the Methodology for Analytical Determination of the Tool Surface Profile

According to a developed methodology for analytical determination of the profile of the tool surface, an example of determining the parameters of a thread-milling cutter for cutting internal threads has been solved. The tool surface of a thread-milling cutter M10 × 1, with which an internal metric thread M20 × 1 is formed, is determined. Table 2 shows the results of solving the problem.

Table 2. Profile parameters of the tool surface of a thread-milling cutter when forming an internal thread M20 × 1.

Name of the Parameter	Designation	Value
	Thread Parameters M20 × 1	
Thread pitch	P, mm	1
Outer diameter	D, mm	20
Medium diameter	D_2, mm	19.350
Inner diameter	D_1, mm	18.917

Table 2. *Cont.*

Name of the Parameter	Designation		Value	
Profile angle	α, °		60	
Number of thread passes	n		1	
Tool Parameters M10 × 1				
Outer diameter of the milling part of the thread-milling cutter	d_0, mm		8.417	
Tool surface profile parameters				
Tool surface center trajectory radius	A_0, mm		5.791	
Parameter of the Archimedean spiral	a		0.275664	
Points from the tool surface profile		point 1	point 3	point 5
Current radius vector of the spiral	R_i, mm	10	9.6752	9.4587
Spiral rotation angle	φ_i, rad	36.2760	35.0979	34.3125
Abscissa of point i	x_i, mm	1.4714	−8.2967	−9.1762
Ordinate of point i	y_i, mm	−9.8912	−4.9775	2.2945
Angular coefficient of the normal in point i	k	−8.2855	0.5618	−0.2812
Parameter	a_1	69.6489	1.3157	1.0791
Parameter	b_1	−38.1083	−0.3552	0.1610
Parameter	c_1	−28.2501	−33.4389	−33.4568
Abscissa of point 1′	$x_{1'}$, mm	0.9667	5.1782	5.4941
Abscissa of point 2′	$x_{2'}$, mm	−0.4196	−4.9083	−5.6433
Ordinate of point 1′	$y_{1'}$, mm	−5.7100	2.5933	−1.8314
Ordinate of point 2′	$y_{2'}$, mm	5.7760	−3.0737	1.3009
Distance between point 1′ and point i	$l_{1'}$, mm	4.2115	15.4560	15.2394
Distance between point 2′ and point i	$l_{2'}$, mm	15.7809	3.8866	3.6700
Radius of the profiled circle	r_{0i}, mm	4.2115	3.8866	3.6700
Applicate of point i	z_i, mm	0.0625	0.2500	0.3750
Distance between point 1′ and point i along the "y" axis of the profiled circle	h_{0i}, mm	0	0.3249	0.5415
Distance between point 1′ and point i along the "z" axis of the profiled circle	z_{0i}, mm	0	0.1875	0.3125
Distance between point 1′ to point i along the "y" axis of the thread	h_i, mm	0	0.3248	0.5413
Difference in profile heights	Δh_i, mm	0	0.0001	0.0002
Tool profile angle	α_{0i}, °	0	59.9805	59.9801
Deviation in profile angle	$\Delta\alpha_{0i}$, °	0	0.0195	0.0199

The calculations were made with the program based on Microsoft Excel 2019. The program contains all the equations from Table 1. The diameter of the thread-milling cutter d_0 and the type of thread that will be formed are entered into the program. After that, the program calculates the parameters of the tool surface.

The designation used for the thread-milling cutter M10 × 1 means M—metric thread, 10—minimum size of the formed thread and 1—thread pitch, P. The same approach is used for the designations of all thread-milling cutters in the publication.

With the values obtained for the points of the tool surface profile, it can be manufactured. Each tooth of the thread-milling cutter must be machined with such a surface. The tool surface obtained by the analytical profiling method is such that when the tool is operated, there is no trimming of the profile of the formed thread.

4. Graphical Determination of the Tool Surface Profile of Thread-Milling Cutters for Milling Internal Threads

4.1. Methodology for Graphical Determination of the Tool Surface Profile

The graphical method for determining the tool surface was developed using the SolidWorks 2022 CAD system [21]. Profiling is performed in the following sequence:

- The main parameters of the formed internal thread D, D2, D1, P and α are defined.
- A three-dimensional model of the internal thread is drawn according to the specified parameters.
- The main design parameters of the milling part of the tool d_0, d_{20}, d_{10}, P_0 and α_0 are defined (Figure 9).
- The radius A_0 is determined according to Equation (29), and the trajectory is drawn that describes the center of the milling part of the tool during the milling process,

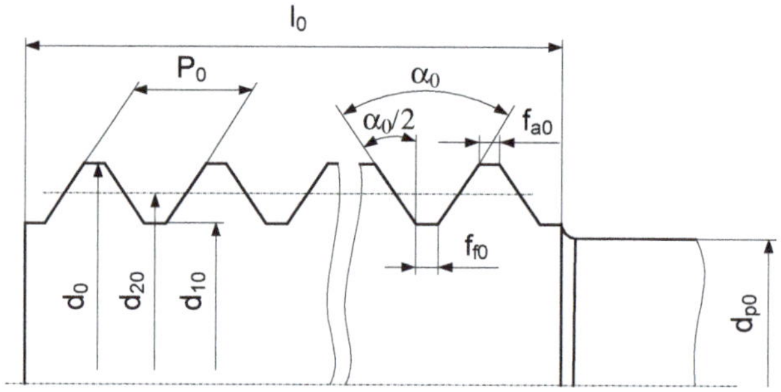

Figure 9. Parameters of the milling part of the thread-milling cutter.

$$A_0 = \frac{D - d_0}{2},\ mm \tag{29}$$

where D is the outer diameter of the formed thread in mm; d_0 is the outer diameter of the milling part of the tool in mm.

- Secant planes intersecting the theoretical profile of the thread with height H from point 1 to point i at distances from z_1 to z_i along the axis of the thread are built (Figure 10).

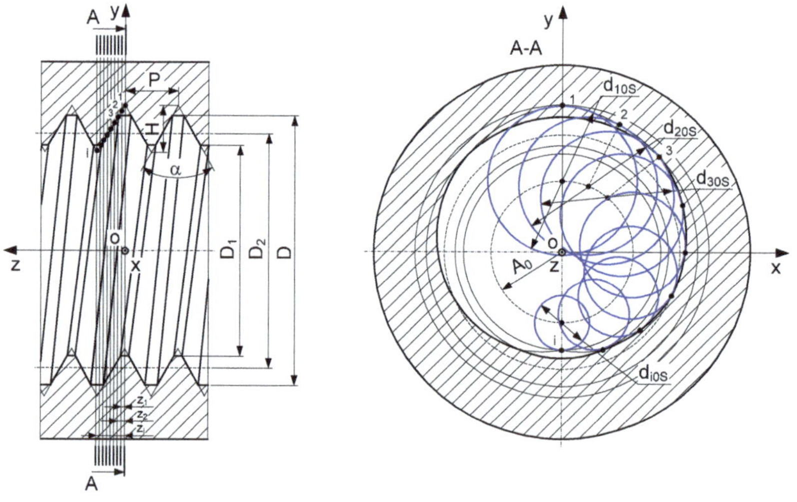

Figure 10. Determining the diameters of the circles forming the tool surface.

The larger number of secant planes determines the larger number of points to describe the profile of the tool surface. This increases its accuracy.

- In each secant plane, circles with diameters d_{10S}, d_{20S}, d_{30S}, ... d_{i0S} are drawn from the tool surface, which are tangent at points 1, 2, 3, ... i of the Archimedean spiral and have a center lying on the circle with radius A_0 (Figure 10).

In Figure 10 these circles are drawn only in section A-A; therefore, they are rotated by a certain angle ε_i proportional to the distances z_i and determined by Equation (2). The circles defining the tool surface profile are shown in Figure 11.

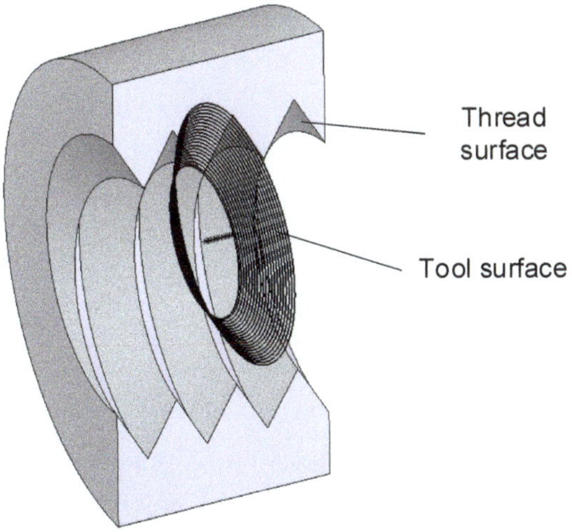

Figure 11. Profiled tool surface.

- The coordinates of points 1, 2, 3, ... i in the "Oxyz" coordinate system are determined, which are necessary for constructing the tool surface and are obtained in the different cross-sections corresponding to the diameters d_{10S}, d_{20S}, d_{30S}, ... d_{i0S} (Figure 12).

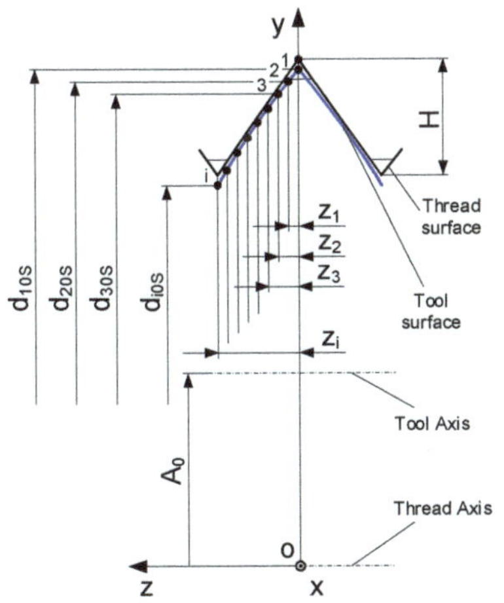

Figure 12. Forming of the theoretical tool surface of the tool.

- The profile of the tool surface is drawn by connecting points 1, 2, 3, ... i with a spline line (Figure 12).

After connecting the points with a spline, the theoretical tool surface of the milling part of the tool is obtained that does not trim the thread profile. The resulting profile has a convex shape.

- The coordinates of the points on the tool surface corresponding to the main diameters of the tool profile d_0, d_{20} and d_{10} are determined (Figure 13).

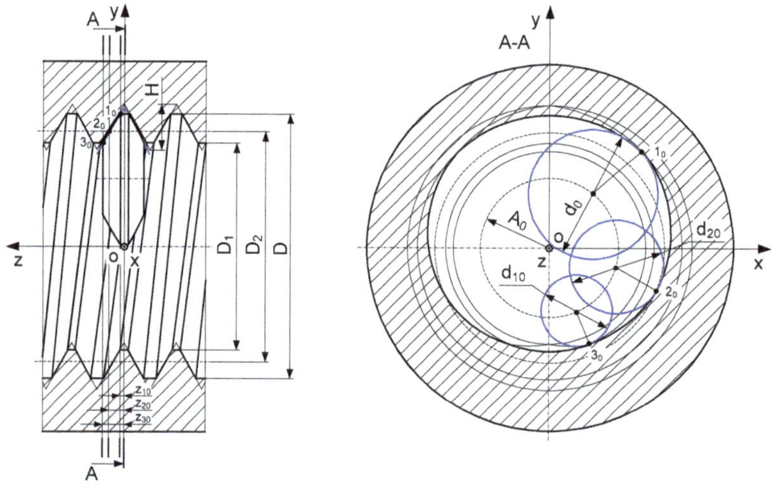

Figure 13. Determining the points on the tool surface corresponding to the main diameters of the tool profile.

With a known radius A_0 of the circle that describes the milling part, the positions of the tool diameters d_0, d_{20} and d_{10} are determined in secant planes through the tool surface obtained by profiling at points 1_0, 2_0 and 3_0. The points are located at distances z_{10}, z_{20} and z_{30}, corresponding to the location of the diameters of the internal thread D, D_2 and D_1 (Figure 13).

- The tool surface profile obtained after profiling is replaced with straight lines (Figure 14).

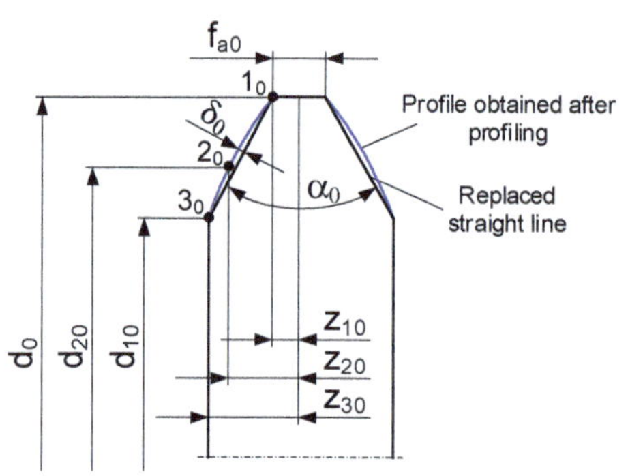

Figure 14. Profile of the surface of the tool obtained after profiling.

The defined profile uses curvilinear forming. Curves are replaced by straight lines from point 1_0 to point 3_0 (Figure 14). As a result of the replacement, the deviation of the profile δ_0 is obtained. The profile of the new tool surface is obtained with a profile angle

$\alpha_0 < \alpha$ as well as a width at the tip $f_{a0} < f_f$, where f_f is the width of the groove of the internal thread.

The size of the error δ_0 is influenced by the pitch P of the formed thread, the outer diameter of the milling part d_0 and the ratio between the diameters of the milling part of the tool and the outer diameter of the thread d_0/D, that is, the maximum diameter of the internal thread that can be machined with a given diameter of the milling part.

Figure 15 shows the profiling of the tool surface of a thread-milling cutter M10 × 1 using the CAD system SolidWorks 2022. The results of the profiling can be seen in Section 5.2. The graphical profiling of the tool surface of milling cutters for forming threads M10 × 1, M16 × 1.5 and M20 × 20 was made. The methodology described in Figure 15 was used for profiling.

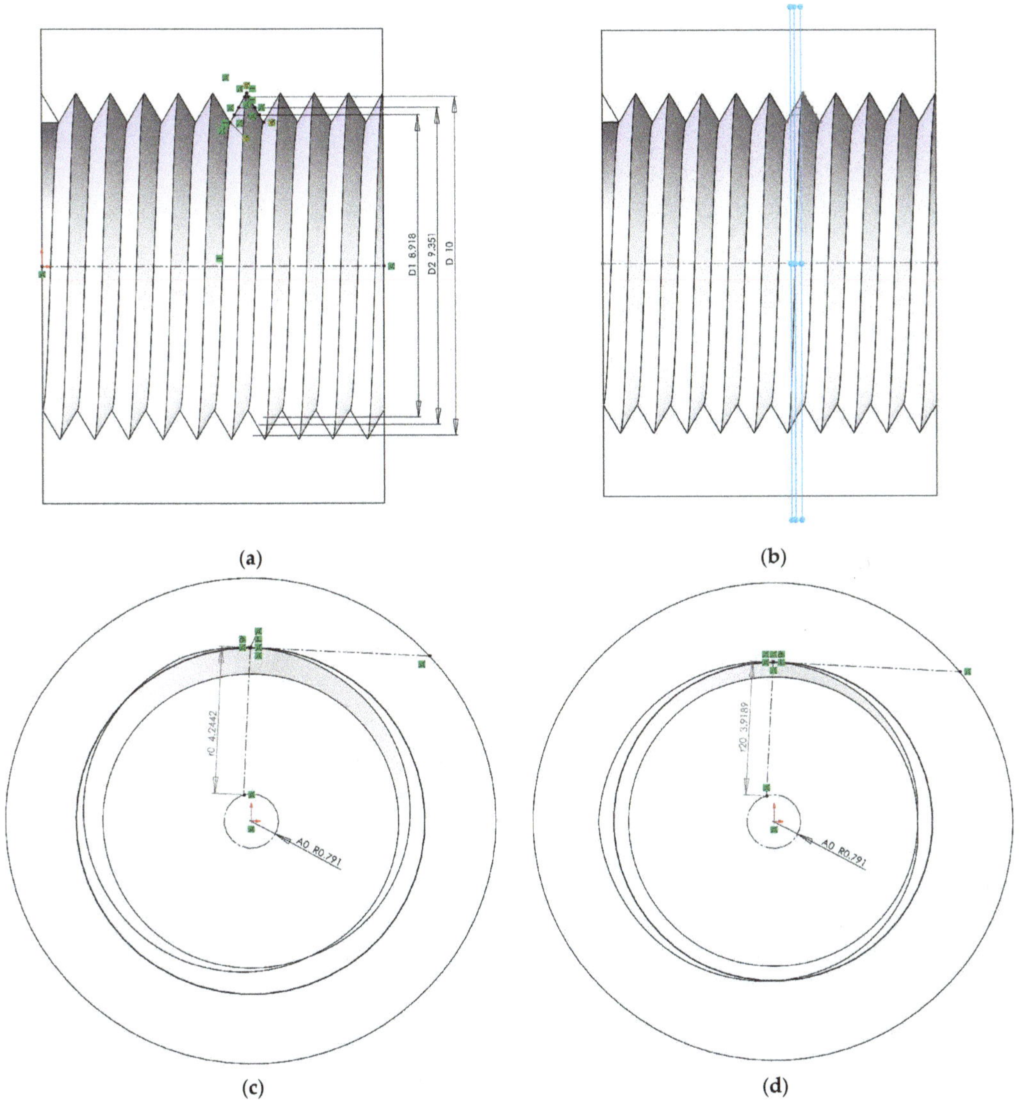

Figure 15. *Cont.*

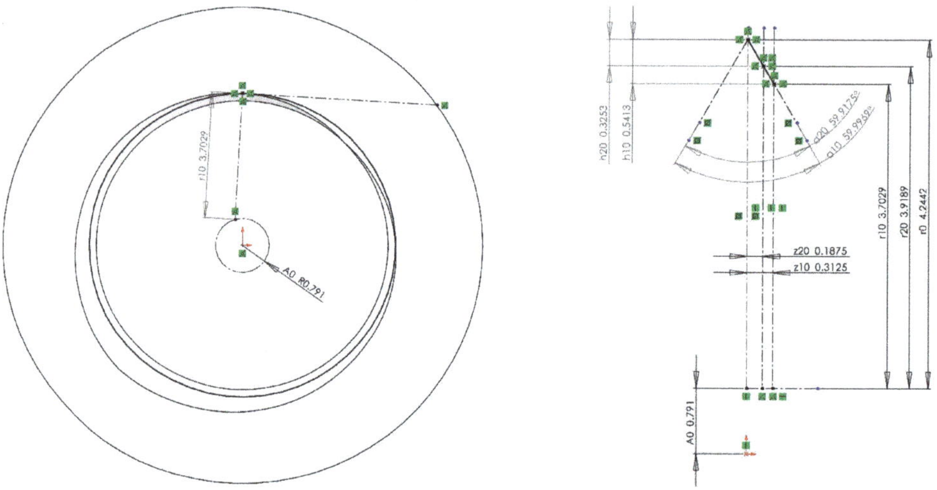

(e) (f)

Figure 15. Graphical profiling of an M10 × 1 thread-milling cutter with SolidWorks 2022: (**a**) Stage 1; (**b**) Stage 2; (**c**) Stage 3; (**d**) Stage 4; (**e**) Stage 5; (**f**) Stage 6.

Profiling with the SolidWorks 2022 CAD system includes the following stages:

- A three-dimensional model of a part with an internal thread M10 × 1 is drawn.
- In a plane built in the cross-section, the profile of the internal thread with diameters D, D_2 and D_1 is drawn. In this way, the position of points 1, 2 and 3 is defined (Figure 15a).
- Planes are built in the axial section passing through points 1, 2 and 3 (Figure 15b).
- In the plane passing through point 1, a circle with radius r_0 is drawn tangent to a point of the Archimedean spiral and with a center lying on the circle with radius A_0 (Figure 15c). The circle corresponds to the diameter of the instrument d_0.
- In the plane passing through point 2, a circle with radius r_{20} is drawn tangent to a point of the Archimedean spiral and with a center lying on the circle with radius A_0 (Figure 15d). The circle corresponds to the diameter of the instrument d_{20}.
- In the plane passing through point 3, a circle with radius r_{10} is drawn tangent to a point of the Archimedean spiral and with a center lying on the circle with radius A_0 (Figure 15e). The circle corresponds to the diameter of the instrument d_{10}.
- A new sketch is built on which the resulting tangent circles with radii r_0, r_{20} and r_{10} are drawn, which form the profile of the tool surface in cross-section (Figure 15f). From the sketch, information is obtained about the profile height h_{0i} and the angle of the tool surface profile α_{0i}.

4.2. Graphical Determination of the Influence of the Rake Angle and the Height of the Backing off on the Tool Surface

The profile of the tool surface is influenced by the rake angle γ_0 and the backing off K_0. They lead to a tooth profile that is different from that in the axial section, i.e., a profile with a different height h_0 and a different profile angle α_0. The values of the rake angle γ_0 depend on the mechanical characteristics of the material being processed and are in the range $0 \div 10°$, and the height of the backing off K_0 determines the values of the back angle at the top points.

The graphical method for determining the influence γ_0 and K_0 was developed using the CAD system SolidWorks 2022 [21]. The profile of the tool surface obtained by analytical or graphical profiling methods is taken as a starting point.

The graphical method is performed in the following sequence:

- The coordinates of the points on the tool surface corresponding to the main diameters of the tool profile d_0, d_{20} and d_{10} are determined (Figure 16).

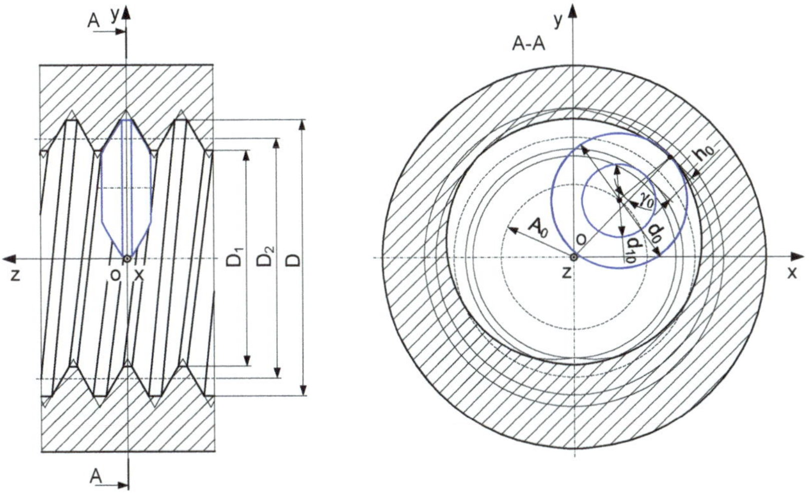

Figure 16. Defining the main design parameters of the tool.

With a known radius of the circle A_0, which describes the milling part, the position of the tool diameters d_0, d_{20} and d_{10} is determined in secant planes through the tool surface obtained by profiling. The secant planes are located at distances corresponding to the location of the internal thread diameters D, D_2 and D_1.

- A secant plane A-A is constructed at a point corresponding to the outer diameter d_0 of the milling part of the tool (Figure 16).
- Graphical builds are made in the secant plane at different rake angles, varying from γ_0 to γ_{0i}, to determine the heights of the profile along the front surface, varying from h_0 to h_{0i} (Figure 17).

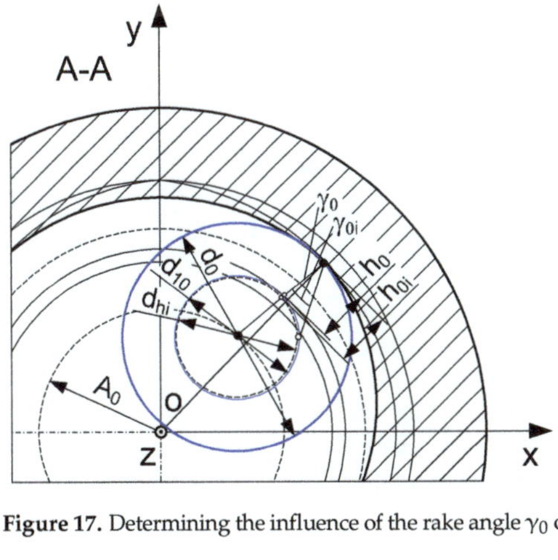

Figure 17. Determining the influence of the rake angle γ_0 on the profile height h_0 along the front surface.

As the rake angle γ_0 increases, the profile height also increases, while its width remains constant.

The influence of the height of the backing off K_0 on the profile angle α_0 is expressed in the presence of the rake angle γ_0. Using the graphical builds shown in Figure 17, the height of the backing off K_0 is also added (Figure 18).

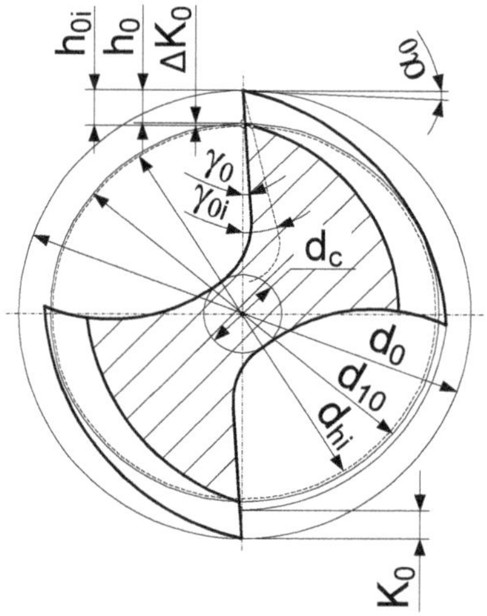

Figure 18. Change in profile height h_0 in the presence of rake angle γ_0 and backing off height K_0.

When the rake angle changes from γ_0 to γ_{0i} and there is a backing off height K_0, the profile height of the tool surface changes from h_0 to h_{0i}, which changes the profile angle from α_0 to α_{0i} and leads to a deviation of the angle $\Delta\alpha_0$ at a constant tooth width b_0 (Figure 19).

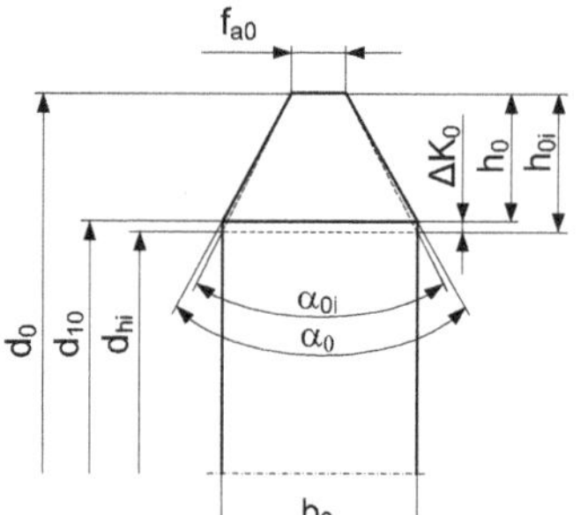

Figure 19. Change in profile height h_0 and profile angle α_0 in the presence of rake angle γ_0 and backing off height K_0.

To form an accurate thread profile, it is necessary to take into account the deviation $\Delta\alpha_0$ of the profile angle:

$$\Delta\alpha_0 = \alpha_0 - \alpha_{0i}, ^\circ \tag{30}$$

The resulting difference ΔK_0 does not affect the accuracy of the thread, since the latter depends on the accuracy of the realization of the center distance.

Figure 20 shows the determination of the influence of the rake angle and the height of the backing off on the tool surface of a M10 × 1 thread-milling cutter using the SolidWorks 2022 CAD system.

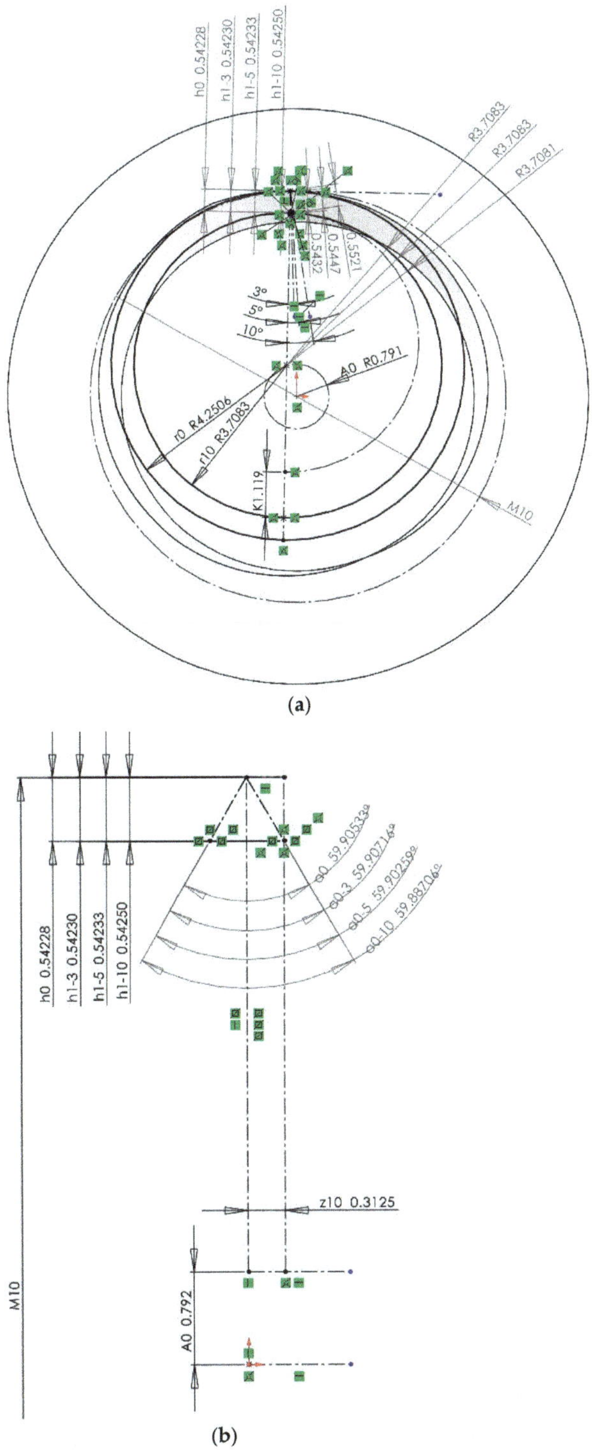

Figure 20. Determining the influence of the rake angle γ_0 and the backing off height K_0 on a M10 $\times$ 1 thread-milling cutter with SolidWorks 2022: (**a**) Stage 1; (**b**) Stage 2.

The results of the profiling can be seen in Section 5.3. The graphical profiling of the tool surface of the milling cutters M10 × 1, M10 × 1.5 and M10 × 2 was performed, which were used to form internal threads with diameters from M10 to M24 with thread pitches P1, P1.5 and P2 mm. For each of the variants, the corresponding three-dimensional models and sketches were developed in SolidWorks 2022. The methodology described in Figure 20 was used for profiling.

Determining the influence of the rake angle γ_0 and the backing off height K_0 with the SolidWorks 2022 CAD system includes the following stages:

- A three-dimensional model of a part with an internal thread M10 × 1 is drawn.
- The coordinates of the points of the tool surface 1 and 3, corresponding to the diameters of the tool profile d_0 and d_{10}, are determined in advance. The coordinates of the points are obtained from the analytical or graphical profiling of the tool surface of the thread-milling cutter.
- A plane is drawn in the axial section passing through point 1 (Figure 20a).
- In the plane with a known radius of the circle A_0, which describes the milling part, circles with radii r_0 and r_{10} are drawn through the points corresponding to the tool diameters d_0 and d_{10}.
- Graphical builds are made in the secant plane, corresponding to the rake angle γ_{0i}—0°, 3°, 5° and 10°. In this way, the heights of the profile along the front surface from h_0 to h_{0i} are determined.
- A curve corresponding to the backing off height K_0 is drawn.
- A new sketch is drawn on which the resulting tool surface is drawn in cross-section (Figure 20b). From the sketch, information is obtained about the profile height h_{0i} and the tool surface profile angle α_{0i} at different values of the rake angle γ_{0i}.

5. Research Methodology and Experimental Results

5.1. Deviations Inherent in the Thread-Milling Scheme with Thread-Milling Cutters Determined by the Analytical Method

Through the developed analytical method for profiling, the deviations of the profile angle of the milling part were determined as $\Delta\alpha_0$. The influence of the change in the diameter of the internal thread D on the profile angle α_0 at different ratios of the diameters of the tool and the thread d_0/D, as well as at a certain pitch of the thread P, was studied. The rake angle of the front surface is γ_0 = 0°. The forming of internal threads of different diameters with thread-milling cutters M10 × 1, M10 × 1.5 and M10 × 2 was studied. Threads M10, M12, M14, M16, M18, M20, M22 and M24 are formed with steps P1, P1.5 and P2. The results obtained are given in Table 3.

Table 3. Deviation of the profile angle $\Delta\alpha_0$ depending on the change in the diameter of the internal thread D and the ratio d_0/D at a certain thread pitch P.

D, mm	d_0, mm	A_0, mm	d_0/D	r_0, mm	r_{20}, mm	r_{10}, mm	α_0, °	$\Delta\alpha_0$, °
				Thread-Milling Cutter M10 × 1				
M10 × 1	8.417	0.792	0.842	4.2506	3.9253	3.7083	59.9125	0.0875
M12 × 1		1.792	0.701	4.2237	3.8986	3.6818	59.9419	0.0581
M14 × 1		2.792	0.601	4.2169	3.8919	3.6752	59.9581	0.0419
M16 × 1		3.792	0.526	4.2140	3.8891	3.6724	59.9683	0.0317
M18 × 1		4.792	0.468	4.2124	3.8875	3.6709	59.9752	0.0248
M20 × 1		5.792	0.421	4.2115	3.8866	3.6700	59.9801	0.0199
M22 × 1		6.792	0.383	4.2109	3.8860	3.6694	59.9836	0.0164
M24 × 1		7.792	0.351	4.2104	3.8856	3.6690	59.9863	0.0137

Table 3. *Cont.*

D, mm	d_0, mm	A_0, mm	d_0/D	r_0, mm	r_{20}, mm	r_{10}, mm	α_0, °	$\Delta\alpha_0$, °
Thread-Milling Cutter M10 × 1.5								
M10 × 1.5	7.876	1.062	0.788	4.0043	3.5152	3.1889	59.7866	0.2134
M12 × 1.5		2.062	0.656	3.9656	3.4772	3.1514	59.8616	0.1384
M14 × 1.5		3.062	0.563	3.9539	3.4658	3.1404	59.9014	0.0986
M16 × 1.5		4.062	0.492	3.9485	3.4606	3.1354	59.9260	0.0740
M18 × 1.5		5.062	0.438	3.9455	3.4578	3.1327	59.9423	0.0577
M20 × 1.5		6.062	0.394	3.9437	3.4561	3.1310	59.9538	0.0462
M22 × 1.5		7.062	0.358	3.9424	3.4549	3.1299	59.9621	0.0379
M24 × 1.5		8.062	0.328	3.9416	3.4541	3.1292	59.9684	0.0316
Thread-Milling Cutter M10 × 2								
M10 × 2 *	7.335	1.333	0.734	3.7552	3.1009	2.6637	59.5885	0.4115
M12 × 2 *		2.333	0.611	3.7078	3.0552	2.6196	59.7389	0.2611
M14 × 2		3.333	0.524	3.6915	3.0397	2.6049	59.8163	0.1837
M16 × 2		4.333	0.458	3.6836	3.0324	2.5981	59.8631	0.1369
M18 × 2		5.333	0.408	3.6791	3.0283	2.5942	59.8939	0.1061
M20 × 2		6.333	0.367	3.6763	3.0257	2.5919	59.9153	0.0847
M22 × 2		7.333	0.333	3.6744	3.0240	2.5903	59.9308	0.0692
M24 × 2		8.333	0.306	3.6730	3.0228	2.5893	59.9424	0.0576

* Threads M10 × 2 and M12 × 2 are not standard. Additional calculations were made for greater completeness of the obtained results.

Figures 21 and 22 present a graphical interpretation of the results obtained in Table 3. Based on the calculations made, the following conclusions can be drawn:

- It was found that with an increase in the diameter of the thread D, at a constant pitch of the thread P and a constant outer diameter of the tool d_0, the deviation of the profile angle of the tool surface $\Delta\alpha_0$ decreases;
- It was found that increasing the pitch of the internal thread P leads to an increase in the deviation of the profile angle of the tool surface $\Delta\alpha_0$;
- It was found that with an increase in the ratio of the diameters of the tool and the thread d_0/D at a constant pitch of the thread P, the deviation of the profile angle of the tool surface $\Delta\alpha_0$ increases.

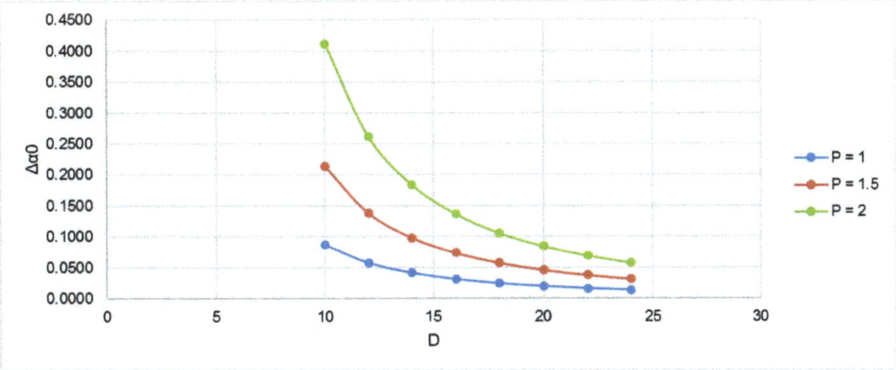

Figure 21. Deviation of the profile angle $\Delta\alpha_0$ depending on the change in the diameter of the internal thread D at a certain thread pitch P.

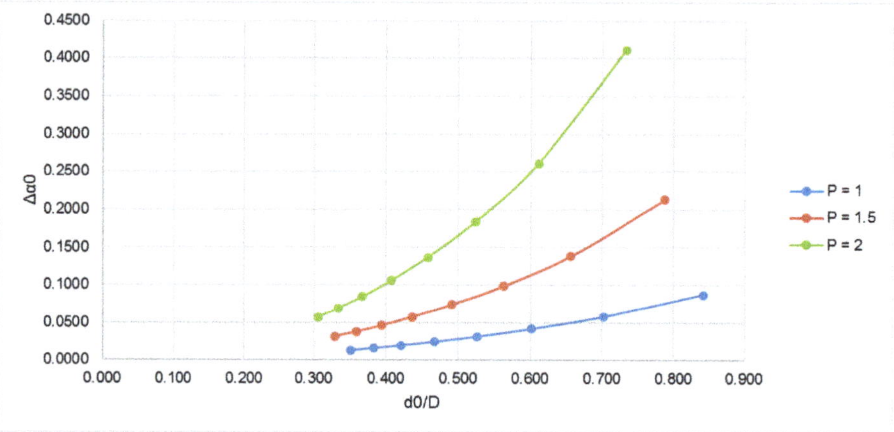

Figure 22. Deviation of the profile angle $\Delta\alpha_0$ depending on the change in the ratio of the diameters of the tool and the thread d_0/D at a certain pitch of the thread P.

5.2. Graphical Verification of the Analytical Profiling Method

In order to compare the results obtained by the analytical or graphical method, the task of determining the profile of the tool surface was solved by both methods when forming different internal threads—thread M10 × 1 with a thread-milling cutter M10 × 1, thread M16 × 1.5 with a thread-milling cutter M10 × 1.5 and thread M20 × 2 with a thread-milling cutter M10 × 2. For the graphical method, the results were obtained from the three-dimensional models and sketches developed for each of the variants with the SolidWorks 2022 CAD system.

For both solutions of the task, the profile is defined in three points—point 1, point 3 and point 5 (Figure 8), with the corresponding radii R, R_2 and R_1. The obtained results are presented in Tables 4–6.

Table 4. Parameters of the profile of the tool surface intended for forming an internal thread M10 × 1, obtained analytically and graphically.

Name of the Parameter	Designation	Value		
Thread parameters M10 × 1				
Thread pitch	P, mm		1	
Outer diameter	D, mm		10	
Medium diameter	D_2, mm		9.350	
Inner diameter	D_1, mm		8.917	
Profile angle	α, °		60	
Number of thread passes	n		1	
Tool Parameters M10 × 1				
Outer diameter of the milling part of the thread-milling cutter	d_0, mm		8.417	
Tool surface profile parameters				
Tool surface center trajectory radius	A_0, mm		0.791	
Points from the tool surface profile		point 1	point 3	point 5
Results obtained using the analytical method				
Radius of the profiled circle	r_{0i}, mm	4.2506	3.9253	3.7083
Distance from point 1′ to point i along the "y" axis of the profiling circle	h_{0i}, mm	0	0.3253	0.5422
Distance from point 1′ to point i along the "z" axis of the profiling circle	z_{0i}, mm	0	0.1875	0.3125
Tool profile angle	α_{0i}, °	0	59.9166	59.9125

Table 4. *Cont.*

Name of the Parameter	Designation	Value		
Deviation in profile angle	$\Delta\alpha_{0i}$, °	0	0.0834	0.0875
Results obtained using the graphical method				
Radius of the profiled circle	r_{0i}, mm	4.2442	3.9189	3.7029
Distance from point 1′ to point i along the "y" axis of the profiling circle	h_{0i}, mm	0	0.3253	0.5413
Distance from point 1′ to point i along the "z" axis of the profiling circle	z_{0i}, mm	0	0.1875	0.3125
Tool profile angle	α_{0i}, °	0	59.9175	59.9969
Deviation in profile angle	$\Delta\alpha_{0i}$, °	0	0.0825	0.0123

Table 5. Parameters of the profile of the tool surface intended for forming an internal thread M16 × 1.5, obtained analytically and graphically.

Name of the Parameter	Designation	Value		
Thread parameters M16 × 1.5				
Thread pitch	P, mm		1.5	
Outer diameter	D, mm		16	
Medium diameter	D_2, mm		15.026	
Inner diameter	D_1, mm		14.376	
Profile angle	α, °		60	
Number of thread passes	n		1	
Tool Parameters M10 × 1.5				
Outer diameter of the milling part of the thread-milling cutter	d_0		7.876 mm	
Tool surface profile parameters				
Tool surface center trajectory radius	A_0, mm		4.062	
Points from the tool surface profile		point 1	point 3	point 5
Results obtained using the analytical method				
Radius of the profiled circle	r_{0i}, mm	3.9485	3.4606	3.1354
Distance from point 1′ to point i along the "y" axis of the profiling circle	h_{0i}, mm	0	0.4878	0.8131
Distance from point 1′ to point i along the "z" axis of the profiling circle	z_{0i}, mm	0	0.2813	0.4688
Tool profile angle	α_{0i}, °	0	59.9292	59.9260
Deviation in profile angle	$\Delta\alpha_{0i}$, °	0	0.0708	0.0740
Results obtained using the graphical method				
Radius of the profiled circle	r_{0i}, mm	3.9468	3.4590	3.1340
Distance from point 1′ to point i along the "y" axis of the profiling circle	h_{0i}, mm	0	0.4878	0.8128
Distance from point 1′ to point i along the "z" axis of the profiling circle	z_{0i}, mm	0	0.2813	0.4688
Tool profile angle	α_{0i}, °	0	59.9328	59.9450
Deviation in profile angle	$\Delta\alpha_{0i}$, °	0	0.0672	0.0550

Table 6. Parameters of the profile of the tool surface intended for forming an internal thread M20 × 2, obtained analytically and graphically.

Name of the Parameter	Designation	Value
Thread parameters M20 × 2		
Thread pitch	P, mm	2
Outer diameter	D, mm	20
Medium diameter	D_2, mm	18.701
Inner diameter	D_1, mm	17.835
Profile angle	α, °	60
Number of thread passes	n	1

Table 6. *Cont.*

Name of the Parameter	Designation	Value		
Tool Parameters M10 × 2				
Outer diameter of the milling part of the thread-milling cutter	d_0, mm		7.335	
Tool surface profile parameters				
Tool surface center trajectory radius	A_0, mm		6.333	
Points from the tool surface profile		point 1	point 3	point 5
Results obtained using the analytical method				
Radius of the profiled circle	r_{0i}, mm	3.6763	3.0257	2.5919
Distance from point 1′ to point i along the "y" axis of the profiling circle	h_{0i}, mm	0	0.6506	1.0844
Distance from point 1′ to point i along the "z" axis of the profiling circle	z_{0i}, mm	0	0.3750	0.6250
Tool profile angle	α_{0i}, °	0	59.9192	59.9153
Deviation in profile angle	$\Delta\alpha_{0i}$, °	0	0.0808	0.0847
Results obtained using the graphical method				
Radius of the profiled circle	r_{0i}, mm	3.6745	3.0240	2.5903
Distance from point 1′ to point i along the "y" axis of the profiling circle	h_{0i}, mm	0	0.6505	1.0842
Distance from point 1′ to point i along the "z" axis of the profiling circle	z_{0i}, mm	0	0.3750	0.6250
Tool profile angle	α_{0i}, mm	0	59.9251	59.9236
Deviation in profile angle	$\Delta\alpha_{0i}$, mm	0	0.0749	0.0764

Based on the calculations made, the following conclusions can be drawn:

- It was found that the values of the profile parameters of the tool surface obtained analytically and graphically differ minimally. The difference obtained is after the third decimal place. The reason for this is the use of input data with values accurate to the fourth decimal place when drawing three-dimensional models, which leads to the accumulation of errors, as well as the permissible minimum deviations that SolidWorks 2022 gives when drawing helical lines and, from there, Archimedean spirals.
- It was found that high-accuracy results are obtained that are fully sufficient for the construction of the tool surface.
- It was proved that the replacement for technological reasons of the curvilinear of the tool surface profile obtained after profiling by analytical or graphical methods with a straight line causes insignificant deviations of the profile δ_0.

5.3. The Influence of the Rake Angle and the Backing Off on the Tool Surface

The deviations of the height of the profile h_0 and of the profile angle of the milling part $\Delta\alpha_0$ were determined by the developed method. The influence of the rake angle γ_0 and the diameter of the internal thread D on the profile angle α_0 was studied with a certain pitch of the thread P and at different ratios of the diameters of the tool and the thread d_0/D. Forming of internal threads from M10 × 1 to M24 × 1 with an M10 × 1 thread-milling cutter, threads from M10 × 1.5 to M24 × 1.5 with a M10 × 1.5 thread-milling cutter, and threads from M10 × 2 to M24 × 2 with an M10 × 2 thread-milling cutter was considered. The results were obtained from the three-dimensional models and sketches developed for each of the variants with the SolidWorks 2022 CAD system. Results are presented in Table 7.

Table 7. Values of the profile height h_0 and the profile angle deviation $\Delta\alpha_0$ depending on the change in the rake angle γ_0 and the thread diameter D for a certain thread pitch P and a constant backing off height K_0.

D, mm	$\gamma_0 = 0°$			$\gamma_0 = 3°$			$\gamma_0 = 5°$			$\gamma_0 = 10°$		
	h_0, mm	α_0, °	$\Delta\alpha_0$, °	h_0, mm	α_0, °	$\Delta\alpha_0$, °	h_0, mm	α_0, °	$\Delta\alpha_0$, °	h_0, mm	α_0, °	$\Delta\alpha_0$, °
Thread-Milling Cutter M10 × 1												
d_0 = 8.417 mm												
K_0 = 1.119 mm												
M10 × 1	0.54228	59.90533	0.09467	0.54230	59.90716	0.09284	0.54233	59.90259	0.09741	0.54250	59.88706	0.11294
M12 × 1	0.54190	59.94192	0.05808	0.54192	59.94009	0.05991	0.54195	59.93734	0.06266	0.54212	59.92179	0.07821
M14 × 1	0.54170	59.96023	0.03977	0.54172	59.95840	0.04160	0.54175	59.95565	0.04435	0.54192	59.94009	0.05991
M16 × 1	0.54160	59.96938	0.03062	0.54162	59.96755	0.03245	0.54166	59.96664	0.03336	0.54182	59.94924	0.05076
M18 × 1	0.54150	59.97854	0.02146	0.54152	59.97671	0.02329	0.54156	59.97305	0.02695	0.54172	59.95840	0.04160
M20 × 1	0.54150	59.97854	0.02146	0.54152	59.97671	0.02329	0.54156	59.97305	0.02695	0.54172	59.95840	0.04160
M22 × 1	0.54150	59.97854	0.02146	0.54152	59.97671	0.02329	0.54156	59.97305	0.02695	0.54172	59.95840	0.04160
M24 × 1	0.54140	59.98771	0.01229	0.54142	59.98587	0.01413	0.54146	59.98221	0.01779	0.54162	59.96755	0.03245
Thread-Milling Cutter M10 × 1.5												
d_0 = 7.876 mm												
K_0 = 1.083 mm												
M10 × 1.5	0.81540	59.79199	0.20801	0.81546	59.78835	0.21165	0.81556	59.78228	0.21772	0.81607	59.75133	0.24867
M12 × 1.5	0.81420	59.86495	0.13505	0.81426	59.86129	0.13871	0.81437	59.85460	0.14540	0.81489	59.82298	0.17702
M14 × 1.5	0.81350	59.90757	0.09243	0.81356	59.90392	0.09608	0.81367	59.89722	0.10278	0.81419	59.86555	0.13445
M16 × 1.5	0.81310	59.93196	0.06804	0.81316	59.92830	0.07170	0.81327	59.92159	0.07841	0.81379	59.88991	0.11009
M18 × 1.5	0.81280	59.95026	0.04974	0.81286	59.94660	0.05340	0.81297	59.93989	0.06011	0.81349	59.90818	0.09182
M20 × 1.5	0.81270	59.95636	0.04364	0.81276	59.95270	0.04730	0.81287	59.94599	0.05401	0.81339	59.91428	0.08572
M22 × 1.5	0.81250	59.96857	0.03143	0.81256	59.96491	0.03509	0.81267	59.95819	0.04181	0.81319	59.92647	0.07353
M24 × 1.5	0.81240	59.97468	0.02532	0.81246	59.97101	0.02899	0.81257	59.96430	0.03570	0.81309	59.93257	0.06743
Thread-Milling Cutter M10 × 2												
d_0 = 7.335 mm												
K_0 = 1.045 mm												
M10 × 2 *	1.09150	59.59147	0.40853	1.09165	59.58468	0.41532	1.09192	59.57246	0.42754	1.09323	59.51324	0.48676
M12 × 2 *	1.08820	59.74120	0.25880	1.08835	59.73438	0.26562	1.08863	59.72165	0.27835	1.08998	59.66035	0.33965
M14 × 2	1.08660	59.81405	0.18595	1.08676	59.80676	0.19324	1.08704	59.79400	0.20600	1.08840	59.73211	0.26789
M16 × 2	1.08550	59.86422	0.13578	1.08566	59.85692	0.14308	1.08594	59.84415	0.15585	1.08731	59.78170	0.21830
M18 × 2	1.08490	59.89163	0.10837	1.08506	59.88432	0.11568	1.08534	59.87150	0.12850	1.08671	59.80904	0.19096
M20 × 2	1.08440	59.91448	0.08552	1.08456	59.90716	0.09284	1.08484	59.89437	0.10563	1.08622	59.83137	0.16863
M22 × 2	1.08410	59.92819	0.07181	1.08426	59.92088	0.07912	1.08454	59.90808	0.09192	1.08592	59.84506	0.15494
M24 × 2	1.08370	59.94649	0.05351	1.08386	59.93917	0.06083	1.08414	59.92637	0.07363	1.08552	59.86331	0.13669

* Threads M10 × 2 and M12 × 2 are not standard. Additional calculations were made for greater completeness of the obtained results.

Figures 23–26 present a graphical interpretation of the results obtained in Table 7. Based on the calculations made, the following conclusions can be drawn:

- It was found that when milling internal threads with tools with a constant outer diameter d_0 and backing off height K_0, with an increase in the rake angle of the teeth γ_0 of the milling part, the height of the profile h_0 and the deviation of the profile angle of the tool surface $\Delta\alpha_0$ increase, and the width of the teeth b_0 remains constant;
- It was found that with an increase in the diameter of the internal thread D at a constant pitch of the thread P and a constant external diameter of the tool d_0, the profile height h_0 and the deviation of the profile angle of the tool surface $\Delta\alpha_0$ decrease;
- It has been proven that in order to obtain an accurate profile of the internal thread, it is necessary to correct the profile of the tool surface by $\Delta\alpha_0$ and ΔK_0.

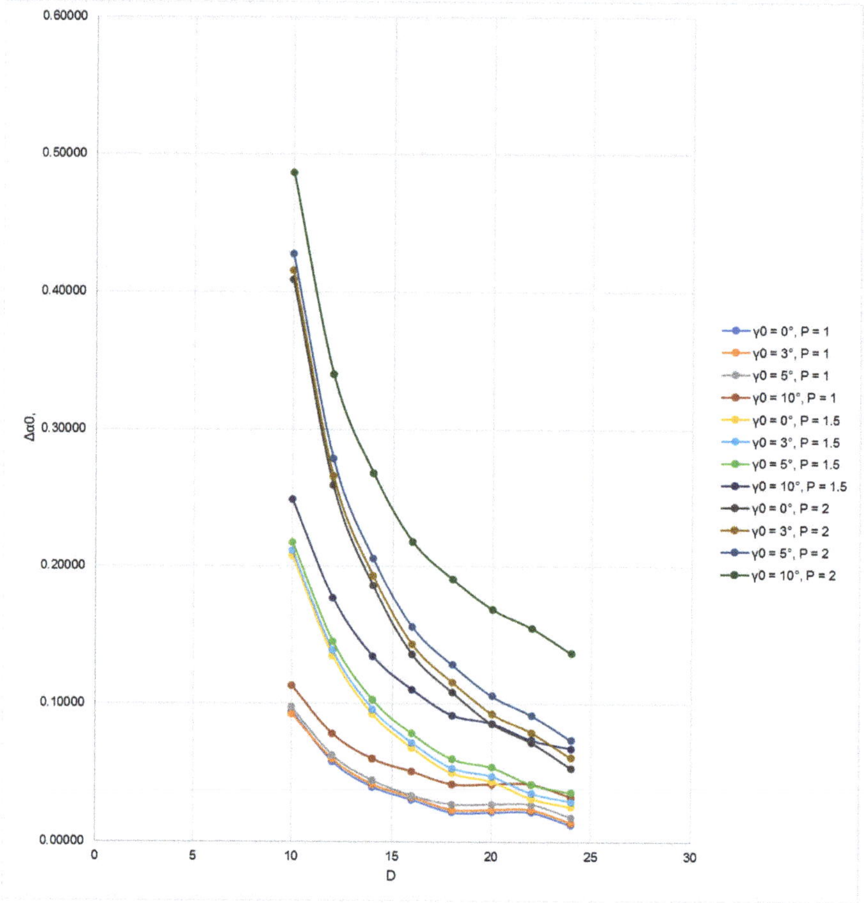

Figure 23. Deviation of the profile angle $\Delta\alpha_0$ depending on the change in the rake angle γ_0, the diameter D and the pitch P of the internal thread.

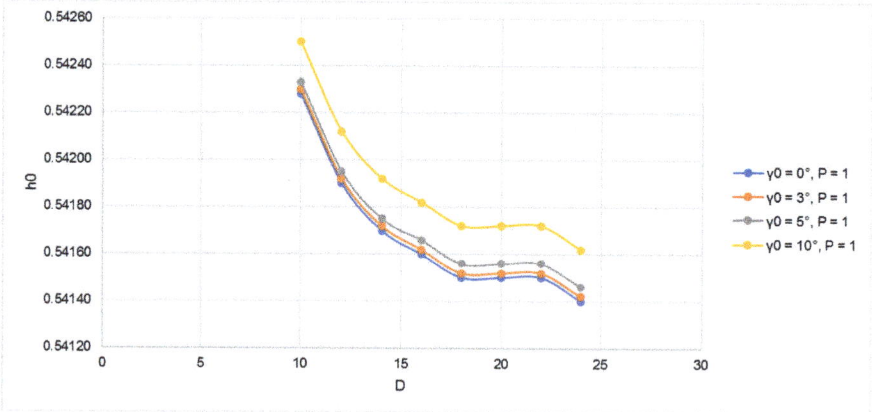

Figure 24. Profile height h_0 depending on the change in the rake angle γ_0 and the diameter of the internal thread D with thread pitch P = 1 mm.

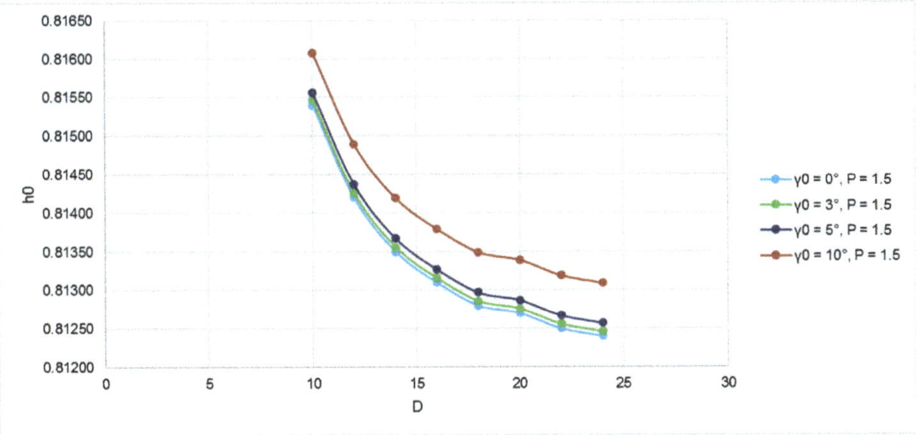

Figure 25. Profile height h_0 depending on the change in the rake angle γ_0 and the diameter of the internal thread D with thread pitch P = 1.5 mm.

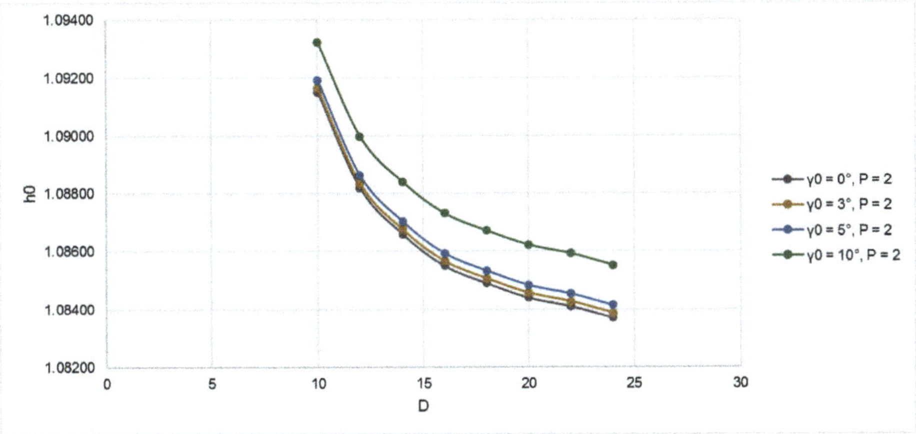

Figure 26. Profile height h_0 depending on the change in the rake angle γ_0 and the diameter of the internal thread D with thread pitch P = 2 mm.

6. Application of Analytical and Graphical Profiling in the Design of Thread-Milling Cutters

The presented methods for profiling the tool surface of internal thread-milling tools have been used in the development of various designs of thread-milling cutters. The following figures present three-dimensional models of tools that are parametrically modeled through configurations. A modeling approach was used in which configurations represent different variants and variations in the models [22].

A specialized methodology was developed to determine the design parameters of the thread-milling cutters. Through the methodology, the equations were determined according to which the parameters of the individual elements of the tool are calculated—milling part, transition part and shank. A Microsoft Excel 2019 program was developed with which the parameters of the thread-milling cutters are determined depending on the internal thread that needs to be formed.

Figure 27 shows the design of a thread-milling cutter designed for forming internal threads with diameters from 6 to 25 mm on CNC machining centers with three or more controlled axes.

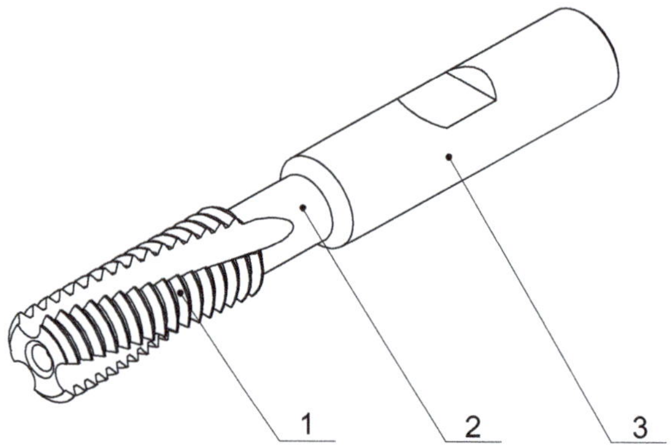

Figure 27. Elements of a thread-milling cutter: 1—milling part, 2—transition part, 3—shank.

The tool consists of a working part and shank 3. The working part consists of a milling part 1 and a transition part 2. The milling part has a length of 2D, where D is the diameter of the formed thread.

Figure 28 shows parametric models of the thread-milling cutter representing different designs with straight and helical flutes. The parameters of the milling part and the shank are parametrically modeled.

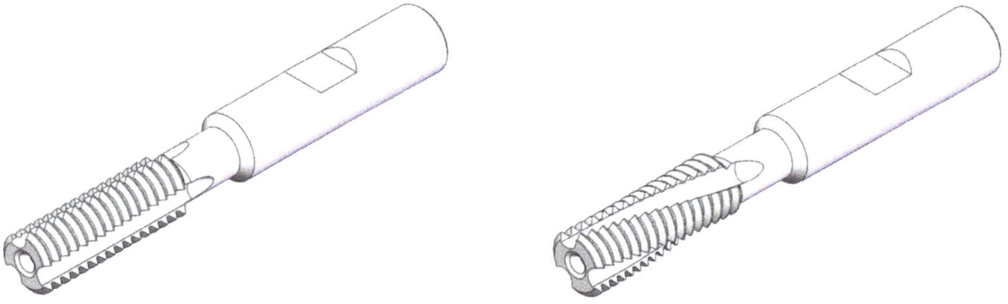

(**a**) (**b**)

Figure 28. Parametric models of the thread-milling cutter: (**a**) Thread-milling cutter with straight flutes. (**b**) Thread-milling cutter with helical flutes.

Design documentation has been prepared for the presented tools for forming internal threads using the SolidWorks 2022 CAD system. The process is automated by using template files for the three-dimensional models and drawings. Attributes (users and systems ones) are filled in the three-dimensional model templates, after which the information from them is visualized in the tables of the drawings [22]. Figures 29 and 30 show the developed drawings of the two models of thread-milling cutters.

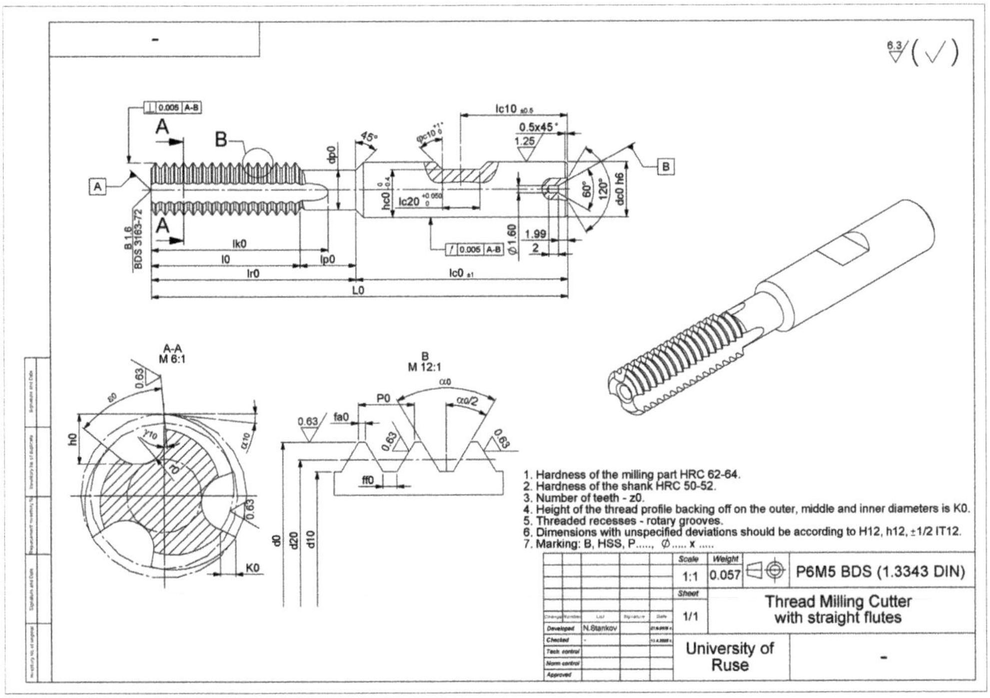

Figure 29. Drawing of a thread-milling cutter with straight flutes.

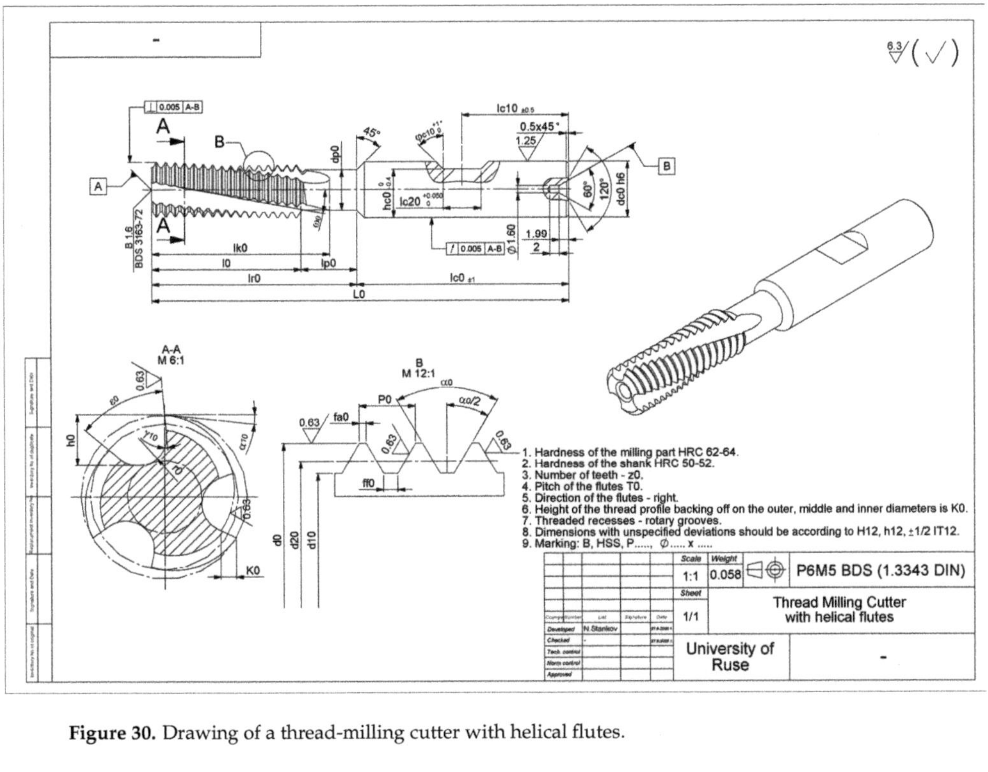

Figure 30. Drawing of a thread-milling cutter with helical flutes.

7. Discussion

The shaping of internal threads by cutting is performed by various methods using tools such as thread-turning tools, taps, thread-milling cutters and thread-cutting heads. Among them, the highest productivity and efficiency is achieved when milling with thread-milling cutters.

The design and manufacture of such tools requires an accurate definition of the tool surface. In this regard, two approaches are presented—analytical and geometric profiling. A comparison of the results shows that both methods give very close values when determining the points of the tool surface.

An analytical method has been developed to determine the profile of the working part of the thread-milling cutters. The developed methodology calculates the points of the profile of each tooth that build the tool surface. For this purpose, a Microsoft Excel 2019-based program has been developed that automatically calculates these points. The obtained parameters allow the design and manufacture of a thread-milling cutter for a specific internal thread or for a group of threads.

In parallel, a graphical method for profiling the tool surface using a CAD system is proposed. This method provides a profile in which the helical surfaces of the formed thread are not trimmed. The implementation using SolidWorks 2022 allows for the assessment of the resulting profile deviations when milling internal threads. The graphical method based on enveloping curves forms the profile by multiple circles that are tangent to an Archimedean spiral in axial section and intersect the thread profile in cross-section. Compared with classical analytical approaches, the proposed graphical methodology allows for a more intuitive verification of the profile and easier application when working with non-standard thread and/or helical profiles.

One of the main advantages of the proposed graphical method is that it is easy to use and implement in CAD software. This makes it accessible even to engineers who may not have much experience in mathematical modeling. Although analytical methods remain more precise from a theoretical perspective, the graphical method is accurate enough for practical applications and helps speed up the design process.

The influence of the design parameters rake angle γ_0 and the backing off K_0 on the profile of the cutting edge has been assessed. The analysis shows that at constant values of the outer diameter d_0 and the backing off K_0, increasing the rake angle γ_0 leads to an increase in the height of the profile h_0 and an increase in the deviation of the profile angle $\Delta\alpha_0$. When the thread diameter D increases, at a constant pitch P and tool diameter d_0, an opposite trend is observed—the height of the profile h_0 and the deviation of the profile angle $\Delta\alpha_0$ decrease. It has also been established that to achieve an accurate profile, it is necessary to correct the profile of the tool surface by $\Delta\alpha_0$ and ΔK_0.

The verification of the results obtained by the analytical method was carried out with the graphical method. Both approaches achieved similar values for the profile parameters. The accuracy of the obtained results is sufficient for the design and subsequent production of a specific thread-milling cutter.

A program has been created in Microsoft Excel 2019 to determine the design parameters of the thread-milling cutter—including the milling part, transition part and the shank. On this basis, designs of thread-milling cutters for internal threads have been developed. Geometric and design parameters have been determined using the analytical and graphical methods. Parametric three-dimensional models have been created in the CAD system and the process of creating design documentation automated using template files. User and system attributes are entered into these files, which are also used in the drawings. Information from the attributes can be used for documentation management through a PDM system as well as by other departments in the company through an ERP system.

However, the methods have their limitations. The results depend on the accuracy of the entered parameters and the CAD environment. The influence of real production factors such as wear, deformations and sharpening errors is not automatically taken into account—these factors are taken into account in the process of the final design of the tool by setting the necessary tolerances, which take into account the type of machining, the type of material being processed, the type of tool material as well as other elements of the cutting mode.

The presented methodology can be used directly in engineering practice, especially in small and medium-sized companies. The methodology is intended for profiling tools for forming all types of threads that have a straight profile. Another advantage is the possibility of the tools having a rake angle γ_0 different from $0°$.

After some additional work, the methodology could be extended to design tools for machining external threads. The same methodology could also be used to profile tools for complex-profile helical surfaces and threads whose profiles are not linear as are the metric threads presented in this publication.

8. Conclusions

The paper proposes and evaluates a graphical profiling method referred to as the Tangent Circles Method. The method for graphical profiling of the tool surface of thread-milling cutters for internal threads is studied. It is based on the use of tangent circles and is implemented in a CAD environment. The method is validated by comparison with results obtained using the analytical method. Simulations with different parameters are also performed. The results show a very good match between the graphical and analytical approaches. The differences in the profile angle are minimal and within the tolerances for real manufacturing.

The graphical method has several practical advantages. It enables quick profile visualization. It is easily performed using a standard CAD system. It requires fewer calculations. It is suitable for designers and technologists who work with tight deadlines and with limited resources. It can also be applied to non-standard tools where analytical approaches become too complex and difficult to apply and require high mathematical knowledge.

The results demonstrate that the Tangent Circles Method is a reliable and effective technique for modeling complex tool geometries using CAD systems. Its ease of use and good accuracy make it a practical alternative to conventional analytical techniques. The method has strong potential for broader application in the design of advanced cutting tools and may be further implemented in automated tool-design systems or extended to support non-standard thread geometries.

However, the method has its inherent limitations. It depends on the accuracy of the CAD software and the correct setting of the parameters. In case of errors in the geometry or insufficient resolution, deviations occur. The method does not take into account the influence of mechanical deformations, wear or sharpening errors.

A positive aspect of the presented work is the analysis of the influence of the rake angle γ_0 and the backing off K_0 on the profile of the tools. It is shown that these parameters have a significant influence on the profile of the cutting edges. By changing the rake angle, the height and inclination of the working surface change. With a combined change in γ_0 and K_0, a visible deviation from the ideal profile is obtained. This shows that profiling should not be considered in isolation. It should be taken into account with the design and technological characteristics of the tool.

The comparative analysis between the analytical and graphical methods confirms the reliability of the tangent circle approach. For different threads and tools, it has been found that the results of the two methods differ by less than one tenth of a degree in the profile angle. This is completely acceptable in design and production.

The proposed graphical method of Tangent Circles is suitable to be used in manufacturing. It is accurate enough for real applications. It is easy to implement and understand. The authors support the idea that this method is also suitable for small and medium-sized companies where the necessary time cannot be allocated and there are no resources or capacity for full analytical modeling. However, if the task has increased accuracy requirements, or when profiling tools is characterized as critical, the proposed method should be used in combination with analytical analysis or with real verification of the results in production.

Author Contributions: Conceptualization, N.S. and A.I.; methodology, A.I. and N.S.; software, N.S.; validation, N.S. and A.I.; formal analysis, N.S.; investigation, N.S.; resources, N.S.; data curation, N.S. and A.I.; writing—original draft preparation, N.S.; writing—review and editing, N.S. and A.I.; visualization, N.S.; supervision, N.S.; project administration, N.S.; funding acquisition, N.S. and A.I. All authors have read and agreed to the published version of the manuscript.

Funding: This study was financed by the European Union-NextGenerationEU, through the National Recovery and Resilience Plan of the Republic of Bulgaria, project No. BG-RRP-2.013-0001.

Institutional Review Board Statement: Not applicable.

Informed Consent Statement: Not applicable.

Data Availability Statement: The raw data supporting the conclusions of this article will be made available by the authors on request.

Conflicts of Interest: The authors declare no conflicts of interest.

References

1. Ivanov, V. *Thread Forming—Methods, Technologies, Tooling, Book I*; "Angel Kanchev" University of Ruse: Ruse, Bulgaria, 2020; ISBN 978-954-712-796-8.
2. Ivanov, V. *Thread Forming—Methods, Technologies, Tooling, Book II*; "Angel Kanchev" University of Ruse: Ruse, Bulgaria, 2020; ISBN 978-954-712-815-6.
3. Liu, C.; Li, Y.; Jiang, X.; Shao, W. Five-axis flank milling tool path generation with curvature continuity and smooth cutting force for pockets. *Chin. J. Aeronaut.* **2020**, *33*, 730–739. [CrossRef]
4. Wang, Y.; Wang, D.; Zhang, S.; Tang, Z.; Wang, L.; Liu, Y. Design and development of a five-axis machine tool with high accuracy, stiffness and efficiency for aero-engine casing manufacturing. *Chin. J. Aeronaut.* **2022**, *35*, 485–496. [CrossRef]
5. Omirou, S.L.; Chasos, C.A. Design and implementation of an innovative canned cycle for variable pitch thread cutting on CNC milling machines. *Int. J. Adv. Manuf. Technol.* **2024**, *130*, 4635–4648. [CrossRef]
6. Kim, J.; Park, J.; Ko, T. End mill design and machining via cutting simulation. *Comput.-Aided Des.* **2008**, *40*, 324–333. [CrossRef]
7. Sun, Y.; Wang, J.; Guo, D.; Zhang, Q. Modeling and numerical simulation for the machining of helical surface profiles on cutting tools. *Int. J. Adv. Manuf. Technol.* **2008**, *36*, 525–534. [CrossRef]
8. Wang, L.; Chen, Z.C. A new CAD/CAM/CAE integration approach to predicting tool deflection of end mills. *Int. J. Adv. Manuf. Technol.* **2014**, *72*, 1677–1686. [CrossRef]
9. Li, G.; Zhou, H.; Jing, X.; Tian, G.; Li, L. Modeling of integral cutting tool grooves using envelope theory and numerical methods. *Int. J. Adv. Manuf. Technol.* **2018**, *98*, 579–591. [CrossRef]
10. Lee, S.; Nestler, A. Simulation-aided Design of Thread Milling Cutter. *Procedia CIRP* **2012**, *1*, 120–125. [CrossRef]
11. Ivanov, V.; Ivanov, A.; Stankov, N. Profiling of disk-shaped thread milling cutters. In *Prospects for Development of Education and Science, Collections of Scientific Articles*; Agricultural University of Plovdiv: Plovdiv, Bulgaria, 2016; pp. 85–91, ISBN 978-617-7214-32-7.
12. Frumuşanu, G.; Teodor, V.; Oancea, N. Ring-tool profiling—Graphical method in CATIA based on Generating trajectories theorem. *IOP Conf. Ser. Mater. Sci. Eng.* **2016**, *161*, 012027. [CrossRef]
13. Avramova, T.; Lefterov, E. Regarding the Strength Calculations of Complex Combined Tools. *IOP Conf. Ser. Mater. Sci. Eng.* **2018**, *400*, 022005. [CrossRef]
14. Avramova, T.; Lefterov, E. Machining of Internal Cylindrical Surfaces with Adjustable Tool. *Procedia Manuf.* **2019**, *32*, 29–35. [CrossRef]

15. Monka, P.; Monkova, K.; Balara, M.; Hloch, S.; Rehor, J.; Andrej, A.; Somsak, M. Design and experimental study of turning tools with linear cutting edges and comparison to commercial tools. *Int. J. Adv. Manuf. Technol.* **2016**, *85*, 2325–2343. [CrossRef]
16. Mikolajczyk, T.; Pimenov, D.; Pruncu, C.; Patra, K.; Latos, H.; Krolczyk, G.; Mia, M.; Klodowski, A.; Gupta, M. Obtaining Various Shapes of Machined Surface Using a Tool with a Multi-Insert Cutting Edge. *Appl. Sci.* **2019**, *9*, 880. [CrossRef]
17. Dragoi, M.V.; Parv, L.; Mija, A.; Oancea, G. A Method to Design Profiled Cutting Tools for Inner Turning. *Symmetry* **2022**, *14*, 2690. [CrossRef]
18. Huang, C.L.; Wei, Y.C. Profile Analysis of Spur Gear Shaping Cutters Based on Sharpened Cutting Edges. *Machines* **2022**, *10*, 484. [CrossRef]
19. Brenci, L.M.; Gurău, L.A. Stratified Characterization of Surface Quality of Beech Processed by Profile Milling. *Appl. Sci.* **2024**, *14*, 129. [CrossRef]
20. Jia, J.; Song, W.; Huang, M.; Zhang, Y.; Yuan, X.; Zhang, X.; Tang, D.; Zhang, T.; Xue, Z.; Jiang, Y. Error control method using crossed axes strategy in the five-axis thread milling process. *Sci. Rep.* **2025**, *15*, 11165. [CrossRef] [PubMed]
21. SolidWorks, 3D Design Software. 2025. Available online: https://www.solidworks.com/ (accessed on 20 June 2025).
22. Stankov, N.; Ivanov, A. Automated Design and Parametric Modeling of Excavator Buckets. *Appl. Sci.* **2025**, *15*, 3389. [CrossRef]

applied sciences

MDPI

Article

Development of Customized Algorithms for the Semi-Automatic Generation of Gradient, Conformal Strut-Based Lattice Structures Using Rhino 8 and Grasshopper: Application and Flexural Testing

Sebastian Geyer [1,2,*], Richard Giefing [2] and Christian Hölzl [3,*]

[1] Institute of Material Science and Technology, Faculty of Mechanical and Industrial Engineering, TU Wien, Getreidemarkt 9, 1060 Wien, Austria
[2] High Tech Manufacturing, Department of Engineering, University of Applied Sciences, Hochschule Campus Wien, Favoritenstraße 226, 1100 Wien, Austria; richard.giefing@alumni.hcw.ac.at
[3] Civil Engineering and Construction Management, Department of Building and Design, University of Applied Sciences, Hochschule Campus Wien, Favoritenstraße 226, 1100 Wien, Austria
* Correspondence: sebastian.geyer@hcw.ac.at (S.G.); christian.hoelzl@hcw.ac.at (C.H.); Tel.: +43-1-606-68-77-2313 (S.G.)

Abstract: In recent years, significant advancements have been made in the field of design for additive manufacturing (DfAM). These advancements have focused on key aspects such as topology optimization (TO), generative design (GD), lattice structures, and AI-based algorithms. This paper presents a methodology for developing custom Grasshopper® algorithms to create strut-based, gradient, and conformal lattice structures. Two test geometries were devised and imported into Grasshopper®, and different lattice structures with varying settings, such as conformity, lattice topology, and strut diameter gradient and cell size gradient, were generated and manufactured. A series of experiments was conducted to assess the impact of input parameters on the formation of lattice structures, their performance in three-point bending tests, and their effect on functionality, applicability, and usability. The experimental investigation yielded clear findings regarding the usability and functionality of the proposed algorithm. However, the findings indicate that although the overall process is usable, improvements are required to streamline the algorithm in order to avoid geometry generation errors and to make it more user-friendly. This approach presents a low-cost, customizable alternative to commercial lattice generation tools, with direct integration in Rhino 8 and Grasshopper®.

Keywords: additive manufacturing (AM); design for additive manufacturing (DfAM); lattice structures; parametric CAD; Rhino 8; Grasshopper

1. Introduction

In a manner analogous to timber or steel truss construction, lattice and support structures are employed in AM to achieve the desired product properties. Support structures facilitate the fabrication of freely suspended, thin, or protruding features with minimal additional material input. Strut-based lattice structures are the typical components of these structures [1]. These elements are incorporated manually by the user or automatically by the relevant program during the component preparation process [2,3]. This method is designed to ensure the most stable and distortion-free production possible [4,5]. Following the manufacturing process, the porous design enables the user to either manually remove or machine away the supports. Processes such as RENA® HIRTISATION® employ specifically developed support structures that can be electro-chemically removed without causing damage to the part.

The implications for design freedom and form freedom of AM are significant factors in the development of suitable parametric CAD solutions [6]. These solutions employ design mechanisms such as topology optimization (TO) [7,8], generative design (GD) [9,10], and lattice structures [11–13]. The objective of implementing these mechanisms is to reduce weight, minimize the number of components in assemblies, and enhance the performance of the individual components. All of the aforementioned mechanisms require substantial computational capabilities. Lattice structures present a particularly challenging aspect in this regard [14–16]. Consequently, the current software tool landscape offers limited capabilities in assisting with the creation of complex lattice structures that incorporate varying conformity or gradients in cell size or strut diameter. Examples of such solutions include nTopology nTop [17], Altair® Sulis™ [18], Altair® Inspire™ [19], Carbon® Design Engine Pro [20], Autodesk Netfabb Ultimate [21], and Metafold [22]. However, these tools are costly, have steep learning curves, and offer limited control over conformal and gradient features, motivating the development of a custom Grasshopper-based solution.

In the context of Design for Additive Manufacturing (DfAM), the utilization of such structures is equally applicable to filling material into volumes of components or in the context of connecting two walls [23,24]. Determining the degree of filling is imperative for minimizing production time, material usage, and weight. The lattice structures generated in the component-volume are generally constant, frequently manifesting cubic base cells that remain consistent irrespective of the CAD-model at hand. The generation of these lattice structures is not based on the shape of the component itself; rather, it is based on external factors, such as the intended stiffness-to-weight ratio, favorable acoustic [25], dampening [26,27], or heat transfer characteristics [28].

Previously mentioned software tools are characterized by their advanced capabilities; however, they are not inherently accessible due to the complexity of their learning curves. Moreover, these tools frequently carry a substantial financial investment [15].

This contribution aims to present an innovative and readily accessible lattice design approach that utilizes the parametric CAD software Rhino 8 (Version 8 SR23, 8.23.25251.13001, 8 September 2025) and the visual programming editor plug-in Grasshopper. This approach is designed to facilitate the development of customized algorithms for the purpose of automatically filling imported CAD geometry with conformal gradient strut-based lattice structures. Apart from the desired input file, the user can choose from a variety of input commands such as voxel size, number of voxels, unit cell topology, conformal or non-conformal design, gradient or constant design, mesh quality and output file type.

2. Materials and Methods

The subsequent sections address the identification and acquisition of all necessary assets, including the requisite software tools and plugins. These sections will also encompass the development of the test series, the specification of the 3-point bending test setup, and the manufacturing of the test specimens.

2.1. Software and Hardware

The custom algorithms were developed in Grasshopper® [29], a visual programming plug-in available in the parametric CAD tool Rhino 8 [30]. Given the range of functions available in Grasshopper®, there are multiple approaches or options for developing an algorithm that can achieve the desired outcome. Ref. [31] proposes an algorithm for the corporation of Voronoi Lattices into CAD models. Other methods of designing custom lattice topologies are shown in refs. [32–34], while [35] showcases a novel approach with a novel type of base cell topology utilizing curved struts as well as pointing out the limitations of available software tools and Grasshopper-based open source algorithms.

In this particular instance, the majority of the functions from the Grasshopper® native library were utilized, with select key components being derived from the plugins Crystallon [36], Dendro [37], IntraLattice [38] and Pufferfish [39].

2.1.1. Crystallon

This plugin is an open-source Grasshopper® toolkit for creating lattice structures. Unlike plugins with independently programmed function blocks, Crystallon v2.0.1 offers editable clusters based on native Grasshopper® components that can be adapted to specific applications and needs by the user. Its modular structure allows for easy integration with other plugins and components. In this work, the components were used to create different cell types and divide geometry into a grid structure based on various input parameters and strut diameter modifications.

2.1.2. Dendro

This is a volumetric modeling plugin based on the OpenVDB library. Dendro v1.0.0. offers several ways to enclose points, curves and meshes as a volumetric data type so that various operations such as smoothing, offsets and unions can then be performed on these volumes. This makes it possible to quickly create mesh bodies with the desired settings and export them in STL format for AM. Certain components use all available processor cores through multithreading, which saves a lot of time in creating the volumes and converting them into a mesh body compared to other native Grasshopper® components or plugins, depending on the hardware used. In this work, the most diverse components for conversion, filtering and merging were used as the final step of lattice structure generation.

2.1.3. IntraLattice

This Grasshopper® plugin is used to generate fixed lattice structures within a 3D design space. Developed in 2015, IntraLattice 0.7.6 served as an open-source, extensible alternative to the commercial solutions available at the time for 3D modeling software for AM applications, a field that was growing rapidly. The plugin was a valuable research tool that served as a platform for breakthroughs in design and optimization. In this work, its components were used to create various unit cells, cell types, and mesh body analysis tools.

2.1.4. Pufferfish

Similarly to a pufferfish, the main focus of the included components of this plugin is enabling shape changes, such as fades, transitions, averages, transformations, and interpolations. The 330 available components in Pufferfish v3.0.0 primarily use parameters and factors as inputs to provide greater control over processes. There are also additional components that simplify common Grasshopper® operations and convert Rhino 8 functions into Grasshopper® function blocks. For this work, geometry-changing components such as "Shell polysurface", but also support components such as "Is Surface Trimmed" were used.

As demonstrated in [15,16], the design of complex lattice structures necessitates substantial computational capabilities. The following hardware was used to compute all predefined lattice topologies for the two test series:

- Processor: AMD Ryzen 9 5900X, 12C/24T, SC@5 GHz, MC@4.5 GHz
- RAM: 64 GB (4x 16 GB Kit), DDR4-3600, CL16-18-18-38
- Graphics: GeForce RTX 3060 OC, 12 GB GDDR6
- HDD: Samsung SSD 980 PRO 2 TB, M.2
- OS: Microsoft Windows 10 Professional (x64) Build 19045.3208

Two test geometries with varying complexity were defined in order to quantify the algorithm's different setting capabilities. As illustrated in Figure 1a the geometry and primary dimensions of the test specimen "arch" are relatively uncomplicated. The test geometry comprises a basic, uniform, and symmetrical arch shape, exhibiting uniform thickness throughout the entire cross-section of the model.

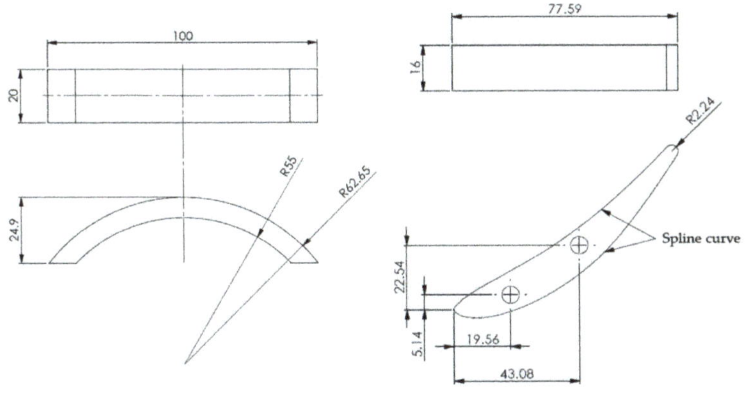

(**a**) (**b**)

Figure 1. Primary dimensions of, test geometry "arch" (**a**) and "wing profile insert" (**b**).

The more complex dimensions of the test geometry "wing profile insert" are shown in Figure 1b. This model is part of a more complex assembly of a formula student race car's rear wing assembly.

2.2. Test Series

Two test series were devised each with a different focus on both test components and the depth of the investigation:

- Test series 1: Partial factorial test with test geometry arch
- Test series 2: Open test with test geometry wing insert (use case Formula Student)

Different base cell topologies were used in the development of the algorithm and the formulation of the test specimens for the two test series. As illustrated in Figure 2a,b, the base cell topologies BCCXYZ and FCCXYZ were utilized in test series 1. A variant of the Diamond base cell topology (Figure 2c) is utilized among the two aforementioned topologies in test series 2.

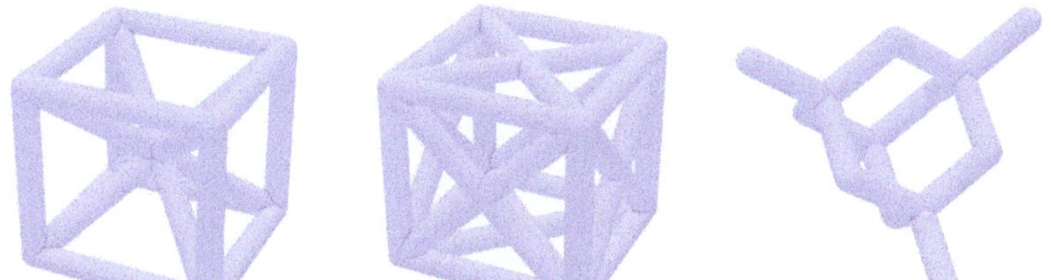

(a) (b) (c)

Figure 2. Base cell topologies used in both the formulation of the algorithm and the subsequent specimens design, (**a**) BZZXYZ, (**b**) FCCXYC, (**c**) Diamond.

Each variation, irrespective of its origin in the test series, was subjected to five iterations of production and subsequent testing. The arithmetic mean, median and confidence intervals were subsequently calculated to obtain meaningful numerical value.

2.2.1. Test Series 1

The parameters investigated exhibit varying degrees of influence on the component's stability (arch and wing profile insert), enabling a good understanding of the parameter's influence on component performance. However, the number and possible gradations of configurable and adjustable parameters that can be used to determine the geometry of the lattice structure and the resulting diversity of variants exceed all test capacities. Consequently, with the help of partial factorial design of experiments, a comprehensive framework was established, encompassing six core factors (A–F), each of which was meticulously evaluated across two distinct levels or settings:

- A (base cell topology): In this series of tests, only BCCXYZ and FCCXYZ were utilized because these cell types are well documented in the literature [27,40,41].
- B (conformity): This factor was instrumental in determining whether the structure conforms to the geometry's surface(s) or not. The surfaces that the lattice structure follows are the upper and lower arch surfaces.
- C (gradient cell size): This parameter was employed to delineate the continuity or gradation of the unit cell size in all principal directions. In principle, further subdivisions could be made based on the axis directions (X, Y, and Z). However, the same defined progression curve was always selected for reasons of simplification.
- D (gradient strut diameter): This factor was employed to ascertain whether the diameter of the struts is uniform or gradually varying. In the case of gradually varying struts, the maximum deviation from the actual diameter was set at 0.2 mm in both the upward and downward directions. Therefore, the smallest value (0.8 mm or 1.22 mm) was designated as the minimum dimension, and the largest value (1.2 mm or 1.62 mm) entered in line E was designated as the maximum dimension. In the event that the bar diameter was uniform, the original value, which was entered in the middle in each case (1 mm or 1.42 mm), was utilized.
- E (strut diameter, when gradient): This parameter served to dictate the selection of the lower or upper diameter range. The upper value is equivalent to twice the cross-sectional area of the lower value.
- F (cell size relative to total volume of design space, when gradient): The average unit cell size, otherwise known as the unit cell volume, is specified as a percentage of the total volume of the design space (input CAD model) in order to ensure the comparability of different component sizes, as well as conformal and non-conformal lattices. The F factor was utilized as the means to control the aforementioned unit cell size. The division of the structure into X, Y, and Z axes was always set so that the calculated percentage value returned in the algorithm corresponds to 0.8 or 1.6%.

Table 1 presents a comprehensive list of these factors, along with their respective levels and the corresponding setting values.

In order to create uniform and comparable unit cell sizes and reasonably manufacturable test specimen, the number of cells and resulting struts that could be implemented for conforming and non-conforming structures for the respective percentage to be achieved was tested in advance. Table 2 presents the resulting values that were utilized for the structures. The mean percentage of correct responses was 0.793% for the lower level and 1.606% for the upper level.

Table 1. Setting table for the study using a partial factorial experimental design.

Factor	Parameter	Settings	
		−	+
A	Base cell topology	BCCXYZ	FCCXYZ
B	Conformity	No	Yes
C	Gradient cell size	No	Yes
D	Gradient strut diameter	No	Yes
E	Strut diameter, when gradient	0.8–1–1.2	1.22–1.42–1.62
F	Cell size relative to total volume of design space, when gradient	0.8%	1.6%

Table 2. Configurations of the unit cell sizes, the numbers in parentheses are chosen when either cell size or strut diameter are gradient.

	0.8%						1.6%					
Factor	Non-Conformal			Conformal			Non-Conformal			Conformal		
Direction	X	Y	Z	X	Y	Z	X	Y	Z	X	Y	Z
Cells	13	4	6(7)	12	4	2	10	3	5(6)	8	3	2
Struts	14	5	7(8)	13	5	3	11	4	6(7)	9	4	3

Achieving optimal consistency and comparability in structural outcomes necessitated meticulous coordination and minimal deviation. This ensured that conformal and non-conformal structures could be reliably and validly compared. The strut diameter exerted a minor influence on this phenomenon, as the available design space experienced a reduction in proportion with larger cross-sections in comparison to the total volume. This was attributable to the corresponding strut radius of the outermost struts.

A full factorial test design with six factors and two levels would result in $2^6 = 64$ different test components. Given the necessity of manufacturing and testing each component multiple times, this approach is not viable given the available resources. Consequently, a partial factorial test plan with six factors and two stages was selected, according to the scheme $2^{(6-2)} = 16$ distinct test components. The test design encompasses six factors and 16 distinct test components. As illustrated in Table 3, the classification of the lower (−) and upper (+) levels is also delineated. [42,43].

Table 3. $2^{(6-2)}$ partial factorial experimental design.

Nr.	A	B	AB	C	AC	AE	E	D	AD	BD	ABD	BF	ACD	F	AF	Test Nr.
1	−	−	+	−	+	+	−	−	+	+	−	+	−	−	+	8
2	+	−	−	−	−	+	+	−	−	+	+	+	+	−	−	7
3	−	+	−	−	+	−	+	−	+	−	+	+	−	+	−	2
4	+	+	+	−	−	−	−	−	−	−	−	+	+	+	+	16
5	−	−	+	+	−	−	+	−	+	+	−	−	+	+	−	13
6	+	−	−	+	+	−	−	−	−	+	+	−	−	+	+	15
7	−	+	−	+	−	+	−	−	+	−	+	−	+	−	+	6
8	+	+	+	+	+	+	+	−	−	−	−	−	−	−	−	14
9	−	−	+	−	+	+	−	+	−	−	+	−	+	+	−	4
10	+	−	−	−	−	+	+	+	+	−	−	−	−	+	+	3
11	−	+	−	−	+	−	+	+	−	+	−	−	+	−	+	11
12	+	+	+	−	−	−	−	+	+	+	+	−	−	−	−	12

Table 3. *Cont.*

Nr.	A	B	AB	C	AC	AE	E	D	AD	BD	ABD	BF	ACD	F	AF	Test Nr.
13	−	−	+	+	−	−	+	+	−	−	+	+	−	−	+	10
14	+	−	−	+	+	−	−	+	+	−	−	+	+	−	−	1
15	−	+	−	+	−	+	−	+	−	+	−	+	−	+	−	9
16	+	+	+	+	+	+	+	+	+	+	+	+	+	+	+	5

The experimental design has a resolution of IV and thus has the following properties according to Siebertz: "Haupteffekte sind mit Dreifachwechselwirkungen vermengt und Zweifachwechselwirkungen untereinander. Das Feld ist geeignet, um Haupteffekte sicher zu bestimmen, Zweifachwechselwirkungen lassen sich jedoch nicht eindeutig zuordnen" [The main effects are interwoven with triple interactions and double interactions with each other. The field is suitable for reliably determining main effects, but double interactions cannot be clearly assigned] [42].

Table 4 presents a comprehensive list of the 16 components and their respective configurations. The utilization of this table facilitates the precise creation of each component in accordance with the specified settings of the algorithm, enabled by the adaptable parameters.

Table 4. Factor settings resulting from experimental design.

Nr.	1	2	3	4	5	6	7	8	9	10	11	12	13	14	15	16
A	−	+	−	+	−	+	−	+	−	+	−	+	−	+	−	+
B	−	−	+	+	−	−	+	+	−	−	+	+	−	−	+	+
C	−	−	−	−	+	+	+	+	−	−	−	−	+	+	+	+
D	−	−	−	−	−	−	−	−	+	+	+	+	+	+	+	+
E	−	+	+	−	+	−	−	+	−	+	+	−	+	−	−	+
F	−	−	+	+	+	+	−	−	+	+	−	−	−	−	+	+

2.2.2. Test Series 2

The test geometry "wing profile insert" was utilized in the open experiments category without the use of an extensive partial factorial experiment design and consisted of a real-world application in the form of a winglet insert for a race car. As illustrated in Figure 1b, the primary dimensions of the solid body (design space) of the winglet profile insert are presented. The distinct geometry configurations were devised by leveraging the insights derived from the initial test series. Given the potential variability in the optimization among different 3D designs and the dynamic nature of the factors in relation to initial tests, the possibility of identifying additional, potentially superior structural variants is possible.

2.3. Test Setup and Manufacturing

All test specimens, that were derived from the two test series and developed from the proposed Grasshopper® algorithm were manufactured out of PA2200 [44] using a Formiga P110 SLS (LPBF) machine by EOS GmbH (Krailling, Germany). The process parameter for the SLS process utilized was the standard "EOS" parameter with a layer thickness of 100 µm. The parts were subsequently processed using a blasting with the use of glass beads. Temperature and humidity conditions were not taken into account.

The load tests were carried out using 3-point bending tests because there is already ample work on tensile and compression tests of lattice structures [27,45,46]. The bending test is also ideal for investigating whether non-conformal or conformal and constant or gradient lattice structures can have a superior force/weight ratio. The 3-point bending tests were conducted using a Testmatic System Z5-X1200 (Nürnberg, Germany) materials testing machine.

The ISO 178:2019 standard was predominantly utilized for the design of the test procedure [47]. In order to ensure the reliability of testing on the test specimen, a specimen holder was developed that deviates from the standard. As illustrated in Figure 3a, the specimen holder and the compression fin affixed within the testing apparatus are depicted from a frontal perspective. Figure 3b presents a visual representation of the specimen holder and the pressure fin positioned at an angle from above, with a test specimen integrated into the assembly. For the lower contact surfaces, a small metal plate was placed between the specimen holder and the test specimen on the left and right. By reducing the friction between the specimen and the holder, the deformation space is not excessively restricted because the support surfaces can slip freely when a load is applied. To mitigate the potential for slippage or forward or backward deviation of the specimens in the X-direction, which could occur with specimens exhibiting imperfect symmetry, the specimens are positioned within a 1 mm deep, transverse groove along the y-axis.

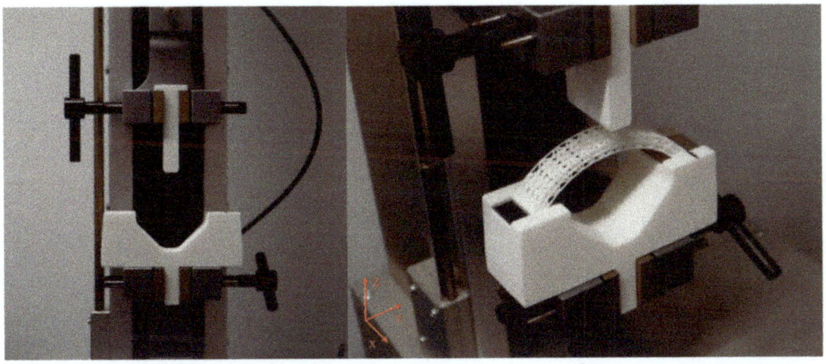

(a) (b)

Figure 3. 3-point bending test setup, (**a**) sample holder and pressure fin clamped in machine, (**b**) test specimen on sample holder.

The configuration of the testing machine was established through a series of preliminary tests conducted prior to the formal test series. Therefore, the subsequent settings shown in Table 5 were established for all test series as a compromise between data quality, the requisite test time, and the various specimens (arch, wing insert) and their specific properties.

Table 5. Settings 3-point bending test.

Parameter	Value
Test speed [mm/min]	10
Max. force [N]	30
Travel [mm]	25
Test method	3 point bending test
y-axis	Force [N]
x-axis	Travel [mm]

3. Results

The next sections delineate the methodologies for identifying and obtaining all required assets, including the required software plugins, development of the algorithm, and testing and validation of the proposed test geometries (partial factorial tests as well as the open tests of both the arch and the wing insert). Additionally, the sections discuss the implications and limitations of the proposed algorithm as well as the used software plugins.

3.1. Algorithm

The fundamental principle guiding the algorithm's design was to ensure its accessibility, even for users with no prior experience with Rhino 8 or Grasshopper®. Consequently, all components and functions necessary for the control of output were arranged in chronological order, with the more complex calculation clusters positioned subsequent to the components that should ideally be defined first, in order to achieve a reliable result. In the subsequent section, the algorithm's excerpts and their respective outcomes are presented for several pivotal components and clusters. This approach aims to offer insight into the algorithm's design structure.

The creation and integration of the unit cells and the lattice structure are based on the components of the Crystallon plugin. The plugin's entire structure is predicated on the use of voxels, which are a spatial unit into which a single unit cell fits. In this context, a voxel serves as the representation of a unit cell within a three-dimensional grid.

The proposed algorithm is divided into eight steps, as illustrated in Figure 4:

1. The process initiates with the definition of the input CAD model as the design space, followed by
2. The subsequent subdivision of this model into voxels.
3. Following the selection of the unit cell topology and its properties conformity and gradients in cell size or strut diameter,
4. The unit cell is integrated into the previously defined voxels.
5. Following the implementation of optional adjustments and fine-tuning such as trimming, addition of material such as shells or closed surfaces, and text labeling,
6. The entirety of the lattice structure and its data set become available for the purpose of documentation.
7. Subsequently, the lattice structure can be meshed and output using the Dendro plugin,
8. Or output (baked) as a 3D model using the Grasshopper function "pipe".

Voxel dimensions were defined via user-adjustable parameters. As illustrated in Figure 5a, the adjustable parameters for the number of divisions into voxels of the input CAD model in each principal direction are displayed. Furthermore, the number of surfaces to which the resulting lattice structure should conform is selected via the input menu. Due to the random numbering of surfaces in each imported CAD model by Rhino 8, users must additionally select the preferred surfaces in Grasshopper® with optical feedback from the Rhino 8 model view. The resulting conformal voxels are shown in Figure 5b, highlighting surfaces 1 and 2 of the input CAD model in the Rhino 8 model view.

Additionally, factors A to F are selected in Grasshopper®. As illustrated in Figure 5c, the setting parameters for factors A to D for test series 2 are clearly defined. Please note that factor E is directly dependent on the selected base cell topology (factor A) as well as on the selection of the strut diameter gradient (factor D). Please note that Factor F was not implemented in test series 2. Figure 5d illustrates the populated voxels of the input CAD model's design space, represented by the struts of the resulting FCCXYZ lattice before the selected gradient strut thickness is added.

To facilitate a range of desirable configurations, including conformal or gradient lattice structures, the algorithm can adopt multiple paths at this stage in the algorithm. These elements are then integrated in accordance with the user's designated selection.

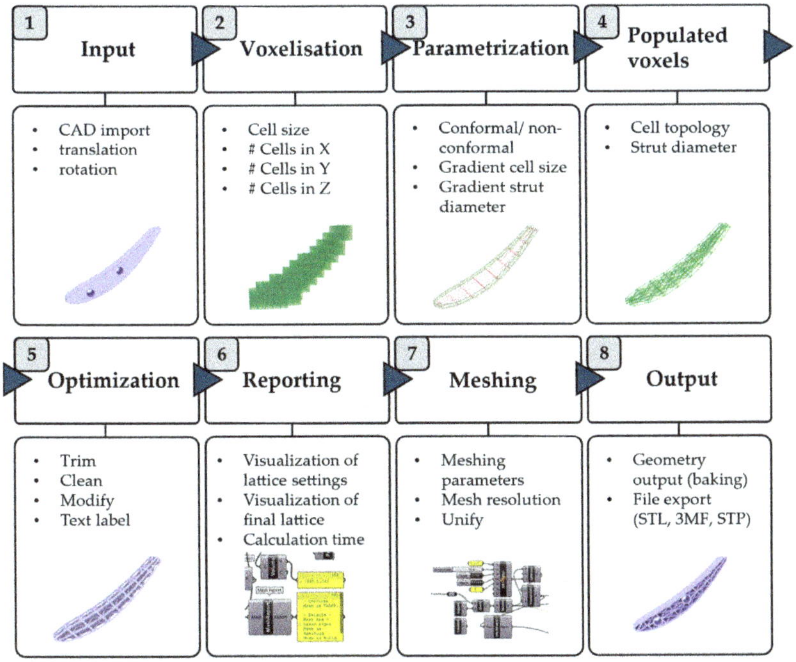

Figure 4. Process description of the proposed algorithm.

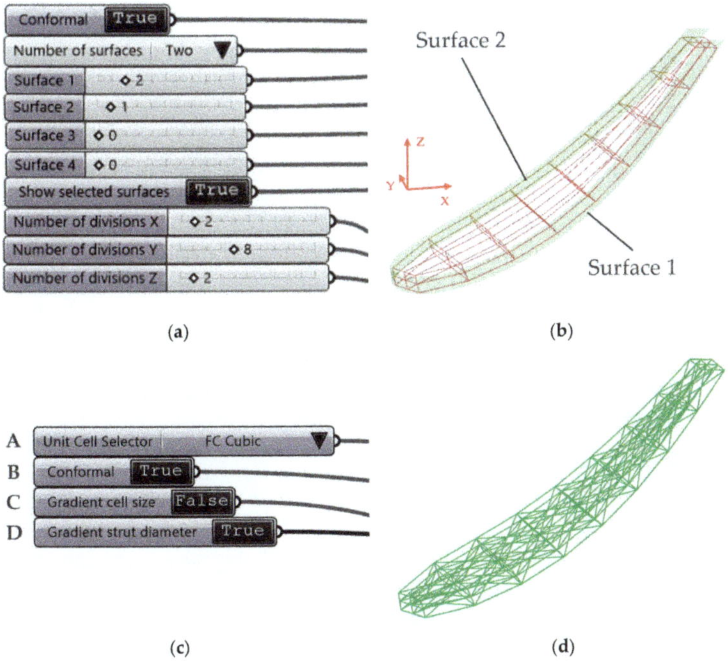

Figure 5. View of the developed algorithm and its visual output, (**a**) adjustable voxel and surface parameters in Grasshopper®, (**b**) resulting voxels and highlighted surfaces the lattice conforms to in Rhino 8, (**c**) selected unit cell parameters in Grasshopper ® for test specimen #3_T4, (**d**) filled voxels with defined FCCXYZ unit cell topology in Rhino 8.

In order to enable uniform gradients for the gradual variation in voxel dimensions, the user can adjust the respective distance between the corner points via the "Graph Mapper-component". The input is invariably a sequence of numbers that delineates the distance between the corner points of the voxels. In the "Graph Mapper" components, this sequence of numbers undergoes a transformation according to the selected curve in the diagram, ensuring its mapping in the sequence of numbers. For example, a sinusoidal curve in the Graph Mapper changes voxel spacing along the x-axis from 5 mm at one end to 2 mm at the other.

Once the topology meets the user's requirements, the final step involves correctly merging the individual 3D bodies and generating the mesh for export, thereby obtaining a part with a high-quality output design. Consequently, there are multiple variants in the algorithm to obtain the final output topology. The appropriate variant and the associated topology type are determined by the respective further processing steps. One of the potential methodologies enables the integration of components from the Dendro plugin.

These geometries undergo a series of processes, including trimming, export, and export verification. In the absence of errors attributable to defective bodies, the part can subsequently be saved as a file. In Figure 6a,c, the input geometry, arc, and wing profile are represented, respectively. In Figure 6b,d, the output models of the lattice structures arch and wing profile inserts are depicted, respectively.

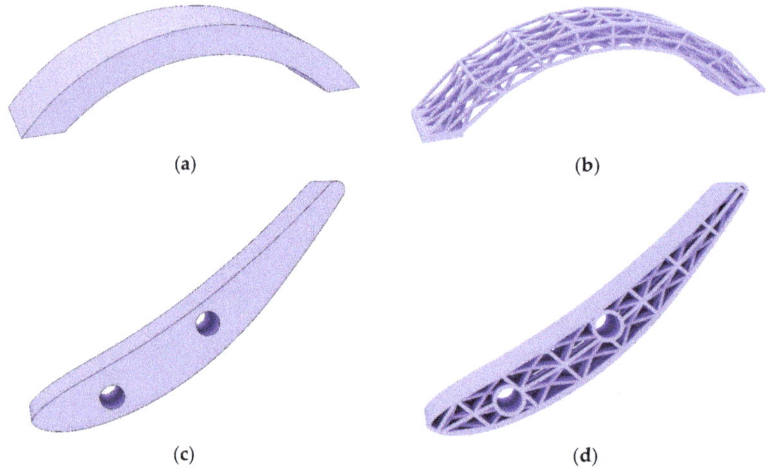

Figure 6. Input and output geometries of the proposed algorithm, (**a**) solid CAD model of the arch, (**b**) conformal BCCXYZ lattice (test part #1_15), (**c**) solid CAD model of the wing insert, (**d**) conformal BCCXYZ lattice (test part #3_T3).

3.2. 3-Point Bending Tests

The following subsections contain the results of the 3-point bending tests, which have been calculated and derived from the test evaluation. The findings presented herein enable the design of lattice structures based on empirical evidence, utilizing designated reference points and corresponding parameter settings. This development eliminates the necessity for conducting simulations and tests to a certain extent.

3.2.1. Test Series 1

The evaluation and results of the partial factorial test series are essential for understanding the individual influencing factors and their effect on component stability. In the process of engineering lattice structures, it is imperative to consider the factors and effects in accordance with the constituent geometry. This approach ensures the optimization of structural stability or weight reduction, thereby maximizing the structure's potential. Table 6 shows the settings used for the 16 individual test specimens of test series 1.

Table 6. Factor settings for the individual test specimen of test series 1.

Nr.	A	B	C	D	E	F
#1_01	BCCXYZ	non-conformal	uniform	uniform	1	0.8
#1_02	FCCXYZ	non-conformal	uniform	uniform	1.42	0.8
#1_03	BCCXYZ	conformal	uniform	uniform	1.42	1.6
#1_04	FCCXYZ	conformal	uniform	uniform	1	1.6
#1_05	BCCXYZ	non-conformal	gradient	uniform	1.42	1.6
#1_06	FCCXYZ	non-conformal	gradient	uniform	1	1.6
#1_07	BCCXYZ	conformal	gradient	uniform	1	0.8
#1_08	FCCXYZ	conformal	gradient	uniform	1.42	0.8
#1_09	BCCXYZ	non-conformal	uniform	gradient	0.8–1–1.2	1.6
#1_10	FCCXYZ	non-conformal	uniform	gradient	1.22–1.42–1.62	1.6
#1_11	BCCXYZ	conformal	uniform	gradient	1.22–1.42–1.62	0.8
#1_12	FCCXYZ	conformal	uniform	gradient	0.8–1–1.2	0.8
#1_13	BCCXYZ	non-conformal	gradient	gradient	1.22–1.42–1.62	0.8
#1_14	FCCXYZ	non-conformal	gradient	gradient	0.8–1–1.2	0.8
#1_15	BCCXYZ	conformal	gradient	gradient	0.8–1–1.2	1.6
#1_16	FCCXYZ	conformal	gradient	gradient	1.22–1.42–1.62	1.6

As illustrated in Figure 7, the mean values of the load curves for the 16 distinct geometries are presented in a force-displacement diagram. The component designated as "#1_08" demonstrated the capacity to withstand the maximum force, while the component identified as "#1_06" exhibited the least resistance. For the test parts designated as #1_02, #1_13, #1_05, #1_14, and #1_01, which are all composed of non-conforming structures, anomalies emerge in comparison to the other parts in the subsequent load curve. Subsequent to the initial ascent of the curve, the force abruptly escalates at a high rate in certain instances. The faulty force curve resulting from excessive deformation is not associated with the actual load-bearing capacity and is therefore ignored.

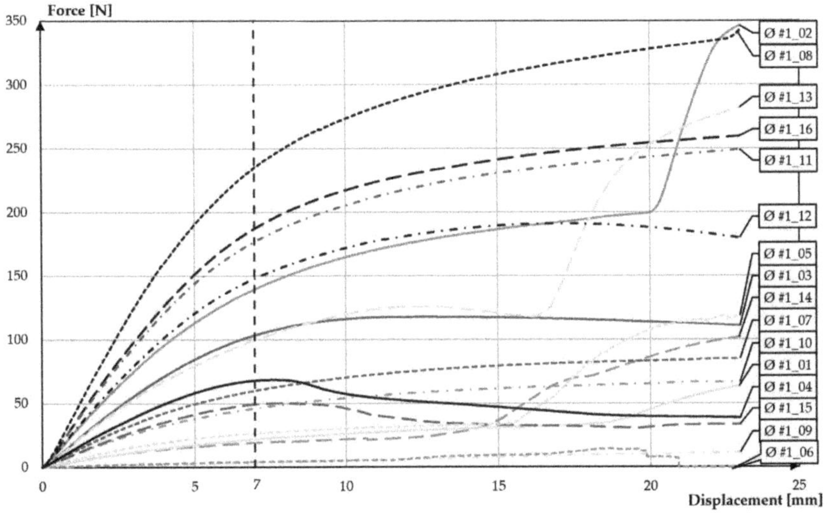

Figure 7. Force-displacement diagram of the averaged measured values of the samples from the partial factorial test plan.

According to ISO 178:2019, the load tests should test up to a bending load ε_f of 5%. According to the formula stipulated in the standard, the maximum deformation for the arch geometry is estimated to be approximately 7 mm (see the vertical dotted line in Figure 7). When this is considered, all anomalies in the load curves are eliminated, rendering them non-decisive for further evaluation. Figure 8a,b show the different behavior of test specimens with different settings under load.

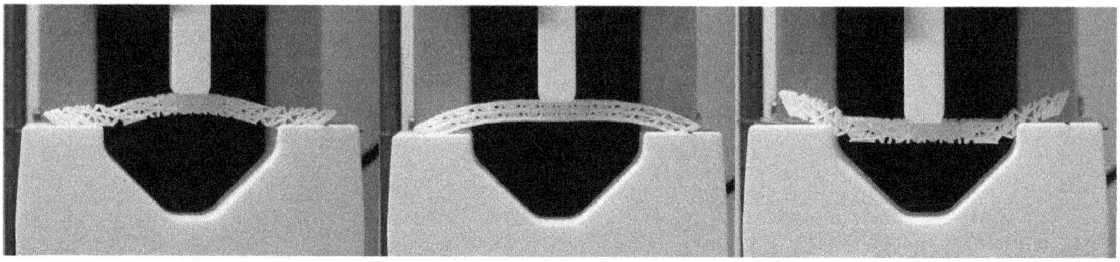

(a) (b) (c)

Figure 8. Comparison between lattice structures under load, (**a**) non-conformal (#1_06), (**b**) conformal (#2_F7) with surrounding closed surface and (**c**) failed non-conformal (#1_06) due to misalignment and binding in the fixture.

The specimen's weak points, due to excessive deformation, cause the contact surfaces to bend upwards, and the middle section of the arch to jam in the specimen holder (Figure 8c). This results in incorrect, excessive force values, as the structure has already failed beforehand.

In order to effectively categorize the components according to their effectiveness in reducing weight while maintaining the same load-bearing capacity, a comparison with the full body is necessary. As illustrated in Figure 9, the respective weight and maximum force achieved by the 16 test parts are compared to the full body as a percentage. It is evident that the components failed to maintain a proportional balance between the reduction in maximum force and the decrease in weight. The component designated as #1_08 exhibited a weight percentage of 63.1% and a maximum force percentage of 46.1%, while component #1_16 demonstrated a weight percentage of 48.5% and a maximum force percentage of 33.9%. These components exhibited optimal performance.

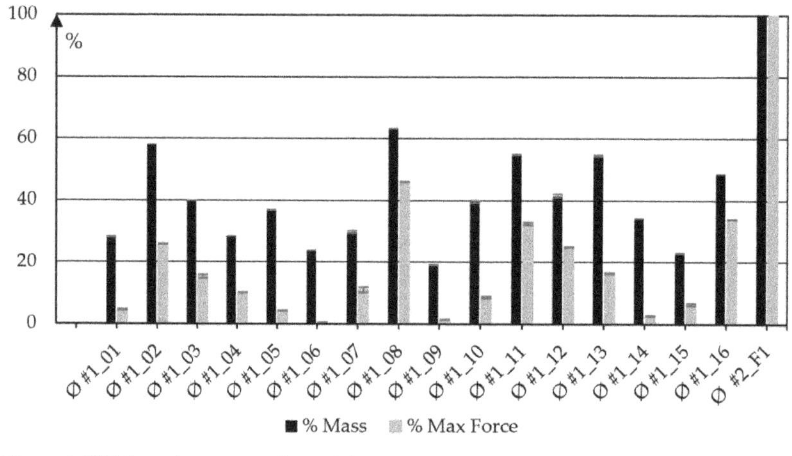

Figure 9. Weight and maximum force of the test parts as a percentage of the arch solid body Ø #2_F1, the confidence intervals are shown in red to emphasize their small margin and make it more visible.

It is imperative to consider factors beyond the fundamental load capacity, namely the force/weight ratio and the impact of individual parameter settings on this ratio. To obtain this, the force/weight ratio was calculated at the maximum force reached in each case. Then, the statistical evaluation of the mean values and effects at this maximum point was conducted. Table 7 presents a comprehensive list of the individual components, along with the respective level settings for the factors and the interactions. In the lower section, the respective mean values are calculated for the upper and lower levels. The subsequent section details the effect calculation, which is derived from the difference between the (+) and (−) mean values. Therefore, the effect signifies the magnitude and direction of the effect that a factor has when either the lower (−) or upper (+) level is integrated into a component. In the event that factor B is utilized as a point of reference, it can be deduced that the incorporation of the upper level (+) into a component, which serves as an indication of the structure's conformity, results in an average enhancement of the force/weight ratio by 15.69 when compared to the utilization of the lower level (−) for the same component. Conversely, in the case of factor F, for example, smaller unit cells (−) provide an average force/weight ratio that is 6.79 times greater than if larger unit cells are used.

Table 7. Evaluation of the maximum force/weight ratio.

Test #	#	A	B	AB	C	AC	AE	E	D	AD	BD	ABD	BF	ACD	F	AF	F/m
8	1	−	−	+	−	+	+	−	−	+	+	−	+	−	−	+	7.9
7	2	+	−	−	−	−	+	+	−	−	+	+	+	+	−	−	21.8
2	3	−	+	−	−	+	−	+	−	+	−	+	+	−	+	−	18.7
16	4	+	+	+	−	−	−	−	−	−	−	−	+	+	+	+	17.3
13	5	−	−	+	+	−	−	+	−	+	+	−	−	+	+	−	5.9
15	6	+	−	−	+	+	−	−	−	−	+	+	−	−	+	+	1.4
6	7	−	+	−	+	−	+	−	−	+	−	+	−	+	−	+	18.1
14	8	+	+	+	+	+	+	+	−	−	−	−	−	−	−	−	35.6
4	9	−	−	+	−	+	+	−	+	−	−	+	−	+	+	−	3.9
3	10	+	−	−	−	−	+	+	+	+	−	−	−	−	+	+	10.7
11	11	−	+	−	−	+	−	+	+	−	+	−	−	+	−	+	28.9
12	12	+	+	+	−	−	−	−	+	+	+	+	−	−	−	−	29.2
10	13	−	−	+	+	−	−	+	+	−	−	+	+	−	−	+	14.7
1	14	+	−	−	+	+	−	−	+	+	−	−	+	+	−	−	4
9	15	−	+	−	+	−	+	−	+	−	+	−	+	−	+	−	14
5	16	+	+	+	+	+	+	+	+	+	+	+	+	+	+	+	34
	MW +	19.2	24.5	18.6	15.9	16.8	18.3	21.3	17.4	16.1	17.9	17.7	16.5	16.7	13.2	16.6	
max.F/m	MW −	14.0	8.8	14.7	17.3	16.4	15	12	15.8	17.2	15.4	15.5	16.7	16.5	20	16.6	
	Eff.	5.24	15.7	3.9	−1.4	0.4	3.3	9.3	1.6	−1.1	2.5	2.2	−0.2	0.2	−6.8	0.00	

The observed range of interaction values extends from 3.85 for AB to 0 for AF. The precise effects of the respective interactions are not enumerated here due to the scope. The force/weight ratio was also determined at 2, 4, 6, 8, 10, and 12 mm so that the influence of the effects can also be seen at different load or, in this case, deformation values. The mean values of the individual stages and the resulting effects were also calculated from this. As illustrated in Figure 10, the resulting effect diagram of factors A–F is based on the force/weight ratio at deformations of 2, 4, 6, 8, 10, and 12 mm, as well as the maximum force. The lower mean value corresponds to the value determined for the negative step, while the upper mean value corresponds to the value determined for the positive step. The effect corresponds to the difference between these values and thus provides an indication of a positive or negative course, as well as the magnitude of the effect. It is evident that factors (C) and (D) exert no significant influence on the force/weight ratio in the context of these investigations. The situation is distinct for factors (A), (B), (E), and (F), some of which exhibit pronounced tendencies. The analysis indicates a favorable outcome in terms of the efficacy of FCCXYZ unit cells in comparison to BCCXYZ cells. The factor (B) demonstrates a clear enhancement in the load-bearing capacity of the components due to the presence of conformal lattice structures. The variation in the parameter (E) indicates that average strut diameters of 1.42 mm compared to 1 mm have a positive effect on component strength. The factor (F) indicates that smaller unit cells are preferable for stability over larger cells.

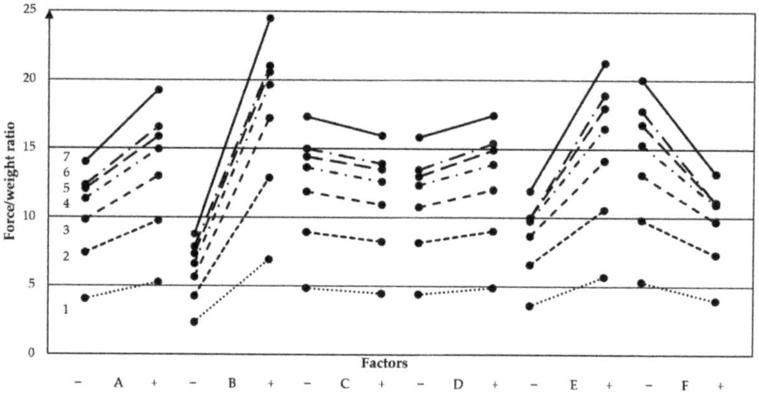

Figure 10. Effect diagram of factors A–F based on the force/weight ratio at different load levels (1) at 2 mm deformation, (2) at 4 mm deformation, (3) at 6 mm deformation, (4) at 8 mm deformation, (5) at 10 mm deformation, (6) at 12 mm deformation, (7) at maximum load/force.

It is imperative to ascertain whether the observed effects emanate from the varied parameter settings or if they are merely a consequence of chance, resulting from extraneous influences. To this end, a normal probability plot can be utilized to ascertain whether the values in a data set are approximately normally distributed or not. Figure 11 displays the standard probability distribution of the primary effects and interaction effects. It is evident that the effects F, E, and B deviate significantly from a normal probability distribution or a random distribution. Consequently, these deviations can be attributed to the change in the factor. Conversely, the majority of the effects are in the positive range, indicating that the "steps/adjustments" have a greater positive effect on the force/weight ratio of the structures than the "- steps/adjustments."

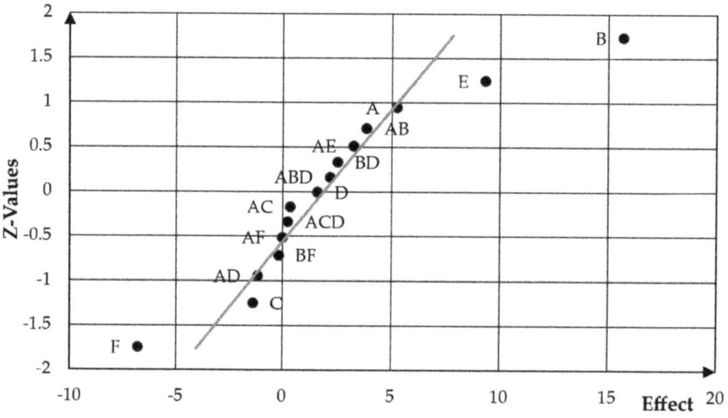

Figure 11. Normal probability for main effects A to Fand interaction effects AB to AF shown in Table 7 with the regression line shown in gray.

3.2.2. Test Series 2

The knowledge acquired from preceding tests was integrated into the development of the geometries of the wing profile inserts. Table 8 shows the settings used for the 6 individual test specimens of test series 2.

Table 8. Factor settings for the individual test specimen of test series 2.

Nr.	A	B	C	D	E	F
#3_T1			Solid body test specimen			
#3_T2			Manually optimized test specimen			
#3_T3	BCCXYZ	conformal	uniform	uniform	1	not implemented
#3_T4	FCCXYZ	conformal	uniform	gradient	1.22–1.42–1.62	not implemented
#3_T5	FCCXYZ	non-conformal	uniform	gradient	1.22–1.42–1.62	not implemented
#3_T6	Diamond	conformal	uniform	gradient	0.8–1–1.2	not implemented

Consequently, it was feasible to enhance the structures in a preliminary iteration to such a degree that certain structures exhibited enhancements in the load tests (Figure 12a) in comparison to the manually optimized original component (Figure 12b).

(**a**) (**b**)

Figure 12. Wing profile insert with different types of internal structures, (**a**) automatically filled design space with conformal FCCXYZ lattice structure (#3_T4), (**b**) manually optimized structure (#3_T2).

As illustrated in Figure 13a, the weight and maximum force of the wing profile inserts during the bending test are expressed as a percentage of the solid body. A thorough examination reveals that none of the components exhibit a higher force/weight ratio in comparison to the full body (Ø #3_T1). For this scenario to be applicable, the gray bar on the right must exceed the black bar on the left. For the component (Ø #3_T4), which has conformal FCCXYZ unit cells with a continuous strut diameter with an average thickness of 1.42 mm, a ratio of almost 1:1 is achieved. It has been determined that, at 58.5% of the full body weight, 58% of the force could be absorbed. This would create a load-bearing alternative corresponding to the lighter weight.

As illustrated in Figure 13b, the weight and maximum force of the wing profile inserts during the bending test are represented as a percentage of the manually optimized and currently utilized component. It is imperative to underscore that component (Ø #3_T4) exhibited the capacity to absorb a 28% greater force compared to a manually optimized component (Ø #3_T2) that possessed an equivalent mass. In a similar manner, the component (Ø #3_T5), which exhibited 86% of the weight, demonstrated the capacity to absorb 28% more force compared to the manually optimized component (Ø #3_T2). This outcome underscores the efficacy of integrating lattice structures within existing components, a concept that holds considerable promise for further refinement through additional optimization of the structural design.

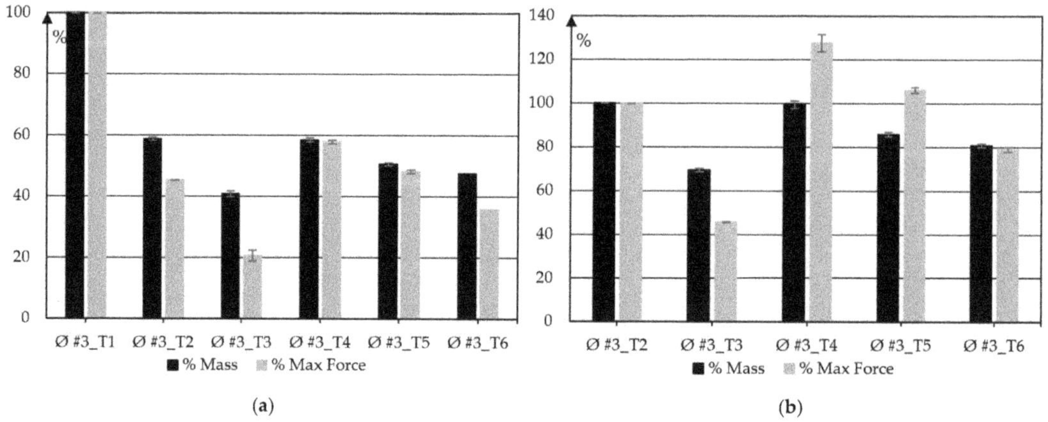

Figure 13. Weight and maximum force of the wing profile inserts, (**a**) as a percentage of the solid body (Ø #3_T1), (**b**) as a percentage of the manually optimized structure (Ø #3_T2), the confidence intervals are shown in red to emphasize their small margin and make it more visible.

4. Discussion

A thorough examination of the advantages, disadvantages, and limitations of Rhino 8 and Grasshopper®, which became evident during the development, implementation, and optimization of the algorithm, as well as a comparison of the newly developed algorithm with existing programs and approaches, is essential to ascertain the novelty and added value of the present study's findings.

4.1. Advantages of Rhino 8 and Grasshopper®

The design possibilities are extensive: The visual programming paradigm is known for being flexible and simple. This makes it possible to implement a wide range of operations. The wide variety of components available allows for easy changes to the lattice parameters. It is thought that users who already know the basics of how to use the system will be able to create or edit data in the right way, although there might be a few small problems along the way.

It has been demonstrated that there are numerous paths that can lead to the same outcome. This dynamic process is a continuous source of novel concepts for the development, adaptation, and optimization of specific functions for various applications.

The software offers a multitude of plugins and extensions, providing users with a wide range of customization options, thereby resulting in a vast array of functional capabilities. The subsequent evolution and incorporation of novel functionalities is dependent not only on the efforts of the developers, but also on the proactive engagement and contribution of the user (open source) community.

A significant amount of literature is available on the subject. The wealth of online resources, including explanatory videos and active forums, ensures that users can acquire the necessary knowledge without reliance on external courses or product support. The diversity of existing application examples also serves as a source of inspiration for one's own implementations, thereby ensuring that the range of tested algorithms and the collective knowledge of the community are in a state of perpetual growth.

The software is characterized by its ability to function seamlessly with a wide array of file formats. The import and export of geometries is not confined to a limited number of file types; rather, it offers a wide range of options that cater to diverse application areas. Consequently, there is no requirement for conversions, which often entail the loss of information. The exportation of STL or STEP files, for instance, is a swift and straightforward process.

With a maximum cost of around EUR 1000 for a permanent license, Rhino 8 and Grasshopper® are competitively priced compared to CAD software on the market, which frequently have much higher price points. There are also more affordable options available for students and individuals pursuing training.

4.2. Disadvantages of Rhino 8 and Grasshopper®

The system's capacity for multithreading is limited. The components in Grasshopper use one CPU core. Additionally, components and function blocks connected in series can only be calculated sequentially. Consequently, the utilization of sophisticated algorithms can result in extended calculation times, during which the program becomes unresponsive, thereby compromising the user experience.

The creation of mesh bodies and "meshes" can lead to a very high RAM requirement for Grasshopper if the level of detail is set too high. Given that Rhino 7 and Grasshopper do not have very high minimum hardware requirements, 16 GB of RAM can quickly become scarce.

In some cases, the reason for a system's functionality or failure can be unclear, particularly when specific parameters have been altered. It is also imperative to exercise caution when combining certain components and the data required by some modules as inputs, ensuring that they do not result in an endless calculation loop.

At sharp corners and edges, notch effects can lead to load peaks, which should therefore be avoided in order not to weaken the structure. However, achieving smooth transitions between individual struts or at junctions can require significant effort or may not be feasible with Grasshopper. While the native Grasshopper feature "Multipipe" provides gradual transitions, it is not fully compatible with gradient strut diameters and gradient cell sizes.

The upper limit of 3D bodies or geometry elements that can be used sensibly with the PC configuration used in this study is approximately 3.000 individual struts. This corresponds to approximately 500 voxels, which may not be sufficient for larger bodies that are to be provided with small unit cells.

4.3. Comparison with Other Software Tools

As previously shown in refs. [15,16], the range of software solutions designed for the creation of 3D CAD models with incorporated lattice structures for AM is now extensive. nTopology's nTop [17] is based on unique software elements that make it possible to exploit the full potential of AM through innovative component design. This makes the software particularly suitable for TO and lattice structure generation. The workflow of the algorithm developed in the course of this work is very similar, in the sense that parts can be imported and lattice structures can be integrated in further steps.

In nTop the processes do not have to be created anew for each design or each component, but function blocks that have been inserted and adapted once can be reused. This approach ensures significant time efficiency by integrating the same operations and associated structures into different parts. The selection of ready-made function blocks, with a variety of setting options, offers easy access to component optimization, which also allows very detailed and specific requests. In contrast to the Grasshopper® algorithm, which was developed from the ground up and now only requires parameter setting, the shown approach involves leveraging existing functions.

In nTop gradients can be embedded in components in a number of ways, making it easier for the user to implement them. For instance, it is possible to insert an image with a grayscale value corresponding to the gradient into the background. This image can then be used as a reference value for adjusting the structure. The program does not offer any simple conversion options for conformal structures. The scope of geometry optimization is defined by the program's functional capabilities and cannot be extended by adding or creating new components, as is possible with Grasshopper®.

Autodesk® Fusion 360 [48], for instance, provides straightforward access to lattice structures for users familiar with the Autodesk® ecosystem. In contrast, Autodesk® Netfabb [21] offers a more in-depth exploration of the realm of gradient lattice structure design.

Altair® Sulis™ 1.12 [18] is a good choice for users looking to enter the field of lattice structures due to its straightforward and user-friendly interface, along with its capacity to swiftly generate satisfactory results in structural integration within a matter of minutes. More complex sequences and combinations can also be implemented with a more advanced understanding of the program.

Carbon® Design Engine Pro [20] is regarded as one of the most advanced tools available on the contemporary market for designing lattice structures. As demonstrated in previous publications, the software tool has the capacity to integrate diverse strut-based lattice structures within a single CAD design space. These structures encompass varied relative density values, distinct gradient values (both cell size and strut diameter), and different conformity levels. The workflow is readily comprehensible and provides effective visual feedback. Furthermore, it enables the incorporation of sensor data, including pressure values measured in millibar, thus providing a multifaceted approach to data integration. The software functions within a browser window on a cloud service, thereby ensuring independence from the computational capabilities of the user's personal hardware.

4.4. Comparison to Prior Research

The investigation of the designed test specimen and the subsequent testing showed typical limitations as well as desirable outcomes when comparing with prior research on this matter. This is particularly relevant when investigating functionally graded lattice structures. Preliminary investigations indicate that for applications requiring high stiffness, strength, or energy absorption (e.g., protective structures, crash absorption, medical implants), it is advantageous to use a smaller unit cell size and sufficiently large strut diameter [49]. However, it is essential to maintain a balance with the manufacturability of the designed structures. If weight is a limiting factor or if material costs are significant, it is essential to maintain strut thickness at a minimum while ensuring the integrity of the design to prevent local instabilities [50]. Functionally graded lattice structures have the potential to meet different requirements within a single component. These structures can be designed with high density in areas where loads are high and with thinner areas where weight can be saved. It is essential to ensure that transitions are as smooth as possible to avoid stress peaks [51]. Manufacturing technology and parameters are critical, and discrepancies often arise between planned values (strut thickness, cell size) and manufactured values [52].

5. Conclusions

This work comprised two major tasks. The initial step involved the creation and implementation of algorithms using Rhino 8 and Grasshopper®. These algorithms have the capacity to semi-automatically integrate lattice structures into chosen 3D designs. The subsequent step entailed the determination of the effects of the individual adjustable parameters on both the algorithms as well as the output lattice designs. The implementation of both of these tasks was successful, as thoroughly delineated in Section 3. The respective conclusions and a general overview and outlook are summarized in the following sections.

5.1. Algorithms

Following a period of familiarization and the implementation of the aforementioned plugins, the comprehensive array of Grasshopper® functions could be utilized to address a wide range of settings, in order to meet user expectations. The resulting algorithms enable the provision of lattice structures tailored to individual preferences. The algorithms function reliably and have been tested with a variety of input 3D models. The quality of the output structures can be adapted accordingly for AM production, and the ability to support a large number of file formats enables free further processing. The tailored algorithms present an alternative solution to existing programs on the market, offering a customizable and expandable range of functions that is not readily available in any other design program today [15,16,53].

Section 4 enumerates the potential extensions to the proposed algorithms and the respective advantages and disadvantages of the Grasshopper® environment. Another potential avenue for exploration pertains to the cross-section geometry of the struts. To illustrate, the implementation of oval struts could facilitate the realization of specific component properties for anisotropic loads and bodies. The implementation of smoothed and adjustable transitions between the individual struts at the crossing points has the potential to enhance the load-bearing capacity of the structures while maintaining a negligible increase in weight.

Furthermore, the proposed algorithms could be made available to the public upon request, which would serve to further enhance performance and user-friendliness.

5.2. 3-Point Bending Tests

The experimental tests and subsequent evaluation of the results provide a clear illustration of the behavior exhibited by the various lattice structures and the respective advantages and disadvantages of the parameters. The failure patterns identified are consistent with the load results. It was demonstrated that, in specific applications, the implementation of conformal lattice structures can yield superior outcomes compared to conventional methods such as increasing strut thickness or reducing unit cell size.

Due to the limited scope of the investigation, it is possible that certain aspects and factors do not fully reflect reality and thus their magnitude. Due to the two-dimensional experimental design, the effects are always linear. However, it is probable that the factors E or F in particular are not linear. To investigate this issue in greater depth, it would be necessary to conduct measurements at multiple stages. A meticulous analysis and representation of all interactions are imperative to draw conclusions about the mutual influence of the factors. The substantial effort required, particularly in the context of experimental designs with numerous parameters, generates the necessary content for an independent study.

The utilization of anisotropic unit cells, exemplified by the FCCZ cell, which were also investigated in refs. [54–57], in accordance with the direction of force holds promise for enhancing the potential for further weight reduction while preserving stability. In addition to the material savings, structures with a superior force-to-weight ratio compared to the corresponding solid body would likely be feasible.

The test subject's small component sizes and the subsequent formation of exceedingly thin struts contribute to the part's remarkable flexibility and "softness." This complicates the distinction between elastic and plastic deformation, and tests rarely result in complete failure, i.e., breakage. This characteristic renders it advantageous when the necessity is for an easily deformable structure or a flexible component. However, it is not useful if the component is required to assume a rigid role. The tests themselves, as well as the evaluation, were also made more difficult because the components easily pushed away or "gave way."

It is posited that the implementation of a test force at a reduced velocity could mitigate the adverse effects, thereby facilitating the acquisition of more precise measurement outcomes. However, this approach entails a substantial increase in the time required for the tests, a circumstance that was not feasible over the course of this work.

Author Contributions: Conceptualization, S.G., R.G. and C.H.; methodology, S.G. and R.G.; investigation, S.G. and R.G.; data curation, S.G., R.G. and C.H.; writing, S.G. and R.G.; visualization, S.G. and R.G.; supervision, C.H. All authors have read and agreed to the published version of the manuscript.

Funding: The research leading to these results was partly funded under the cooperative doctoral program "Digiphot" between Hochschule Campus Wien and TU Wien. Open Access Funding by TU Wien.

Institutional Review Board Statement: Not applicable.

Informed Consent Statement: Not applicable.

Data Availability Statement: The raw data supporting the conclusions of this article will be made available by the authors on request.

Acknowledgments: We would like to express our gratitude to David Nechi for his assistance with the manufacturing and post-processing of the SLS components, which were built by our EOS Formiga P110. I would like to thank Jürgen Stampfl for the supervision of this work and for the supervision of my dissertation at TU Wien. The authors acknowledge TU Wien Bibliothek for financial support and Open Access Funding by TU Wien.

Conflicts of Interest: The authors declare no conflicts of interest.

References

1. Griffiths, L. Tctmagazine. Available online: https://www.tctmagazine.com/additive-manufacturing-3d-printing-news/materialisee-%20stage-support-generation-for-metal-3d-printing/ (accessed on 18 April 2025).
2. Dimopoulos, A.; Salimi, M.; Gan, T.-H.; Chatzakos, P. Support Structures Optimisation for High-Quality Metal Additive Manufacturing with Laser Powder Bed Fusion: A Numerical Simulation Study. *Materials* **2023**, *16*, 7164. [CrossRef]
3. Stanciu Birlescu, A.; Balc, N. Computational Model for Tree-like Fractals Used as Internal Structures for Additive Manufacturing Parts. *Appl. Sci.* **2023**, *13*, 11187. [CrossRef]
4. Subedi, S.C.; Shahba, A.; Thevamaran, M.; Thoma, D.J.; Suresh, K. Towards the optimal design of support structures for laser powder bed fusion-based metal additive manufacturing via thermal equivalent static loads. *Addit. Manuf.* **2022**, *57*, 102956. [CrossRef]
5. Didier, P.; Le Coz, G.; Robin, G.; Lohmuller, P.; Piotrowski, B.; Moufki, A.; Laheurte, P. Consideration of SLM additive manufacturing supports on the stability of flexible structures in finish milling. *Manuf. Process.* **2021**, *62*, 213. [CrossRef]
6. Egan, P.F. Design for Additive Manufacturing: Recent Innovations and Future Directions. *Designs* **2023**, *7*, 83. [CrossRef]
7. Stiefelbauer, A. Application of Biomimetics and Computational Structure Optimization Process for Product Development-Using the Example of an Additive Manufactured Skibinding. Master's Thesis, University of Applied Sciences Vienna-Hochschule Campus Wien. Available online: https://pub.hcw.ac.at/obvfcwmetaonly/content/titleinfo/6612994 (accessed on 22 July 2025).
8. Tyflopoulos, E.; Flem, D.T.; Steinert, M.; Olsen, A. State of the Art of Generative Design and Topology Optimization and Potential Research Needs. NordDesign. 2018. Available online: http://hdl.handle.net/11250/2586059 (accessed on 12 September 2025).
9. General Motors Seat Clamp. Available online: https://www.sculpteo.com/de/3d-lernzentrum/erstellung-einer-druckbaren-3d-datei/generatives-design/ (accessed on 18 July 2025).
10. Szabó, K.; Hegedűs, G. Brief overview of generative design support software. *Des. Mach. Struct.* **2020**, *10*, 123. [CrossRef]
11. Panesar, A.; Abdi, M.; Hickman, D.; Ashcroft, I. Strategies for functionally graded lattice structures derived using topology optimisation for Additive Manufacturing. *Addit. Manuf.* **2018**, *19*, 81–94. [CrossRef]
12. Suzuki, A.; Nakatani, H.; Nakagawa, S.; Kobashi, M.; Tsuji, Y. Architected materials informatics: Construction and application to cellular-structured heat sink optimization. *Acta Mater.* **2025**, *283*, 120557. [CrossRef]
13. Korshunova, N.; Alaimo, G.; Hosseini, S.B.; Carraturo, M.; Reali, A.; Niiranen, J.; Auricchio, F.; Rank, E.; Kollmannsberger, S. Bending behavior of octet-truss lattice structures: Modelling options, numerical characterization and experimental validation. *Mater. Des.* **2021**, *205*, 109693. [CrossRef]

14. Majari, P.; Olvera-Trejo, D.; Estrada-Díaz, J.A.; Elías-Zúñiga, A.; Martinez-Romero, O.; Ramírez-Herrera, C.A.; Perales-Martínez, I.A. Enhanced Lightweight Structures Through Brachistochrone-Inspired Lattice Design. *Polymers* **2025**, *17*, 654. [CrossRef]
15. Geyer, S.; Hölzl, C. Comparison of CAD Software for Designing Cellular Structures for Additive Manufacturing. *Appl. Sci.* **2024**, *14*, 3306. [CrossRef]
16. Geyer, S.; Schwemmer, J.; Hölzl, C. Development and Evaluation of Customized Bike Saddle Pads Using Innovative Design for AM Approaches and Suitable Additive Manufacturing Processes. *Appl. Sci.* **2025**, *15*, 472. [CrossRef]
17. nTopology, Inc. nTop. Available online: https://www.ntop.com/ (accessed on 8 July 2025).
18. Altair Engineering Inc. Sulis. Available online: https://www.altair.com/sulis/ (accessed on 8 July 2025).
19. Altair Engineering Inc. Inspire. Available online: https://altair.com/inspire/ (accessed on 8 July 2025).
20. Carbon Inc. Design Engine Pro. Available online: https://www.carbon3d.com/products/carbon-design-engine (accessed on 8 July 2025).
21. Autodesk Netfabb Ultimate. Available online: https://www.autodesk.com/de/products/netfabb/overview?plc=F360NFP&term=1-YEAR&support=ADVANCED&quantity=1&tab=subscription (accessed on 8 July 2025).
22. Metafold 3D. Metafold. Available online: https://www.metafold3d.com/ (accessed on 8 July 2025).
23. Shields, G. All3DP. Available online: https://all3dp.com/1/3d-printing-lattice-structures-the-ultimate-guide/ (accessed on 19 April 2024).
24. Boschetto, A.; Bottini, L.; Macera, L.; Vatanparast, S. Additive Manufacturing for Lightweighting Satellite Platform. *Appl. Sci.* **2023**, *13*, 2809. [CrossRef]
25. Suárez, L.; Del Mar Espinosa, M. Assessment on the use of additive manufacturing technologies for acoustic applications. *Int. J. Adv. Manuf. Technol.* **2020**, *109*, 2691. [CrossRef]
26. Dong, G.; Tessier, D.; Zhao, Y.F. Design of Shoe Soles Using Lattice Structures Fabricated by Additive Manufacturing. *Proc. Des. Soc. Int. Conf. Eng. Des.* **2019**, *1*, 719. [CrossRef]
27. Harish, A.; Alsaleh, N.A.; Ahmadein, M.; Elfar, A.A.; Djuansjah, J.; Hassanin, H.; El-Sayed, M.A.; Essa, K. Designing Lightweight 3D-Printable Bioinspired Structures for Enhanced Compression and Energy Absorption Properties. *Polymers* **2024**, *16*, 729. [CrossRef] [PubMed]
28. Chen, L.; Zhang, R.; Chu, X.; Liu, H. Multiscale Design of Graded Stochastic Cellular Structures for the Heat Transfer Problem. *Appl. Sci.* **2023**, *13*, 4409. [CrossRef]
29. Grasshopper 3D, Grasshopper. Available online: https://www.grasshopper3d.com/ (accessed on 8 July 2025).
30. Robert McNeel & Associates, Rhino 7. Available online: https://www.rhino3d.com/ (accessed on 8 July 2025).
31. Efstathiadis, A.; Symeonidou, I.; Tsongas, K.; Tzimtzimis, E.K.; Tzetzis, D. Parametric Design and Mechanical Characterization of 3D-Printed PLA Composite Biomimetic Voronoi Lattices Inspired by the Stereom of Sea Urchins. *Compos. Sci.* **2023**, *7*, 3. [CrossRef]
32. Sasagawa, T.; Nimura, N.; Tanaka, M. Flexible lattice structure using curved struts based on body-centered cubic structure. *Addit. Manuf.* **2025**, *103*, 104746. [CrossRef]
33. Ferreiraa, K.; Anwera, N.; Mehdi-Souzani, C. Characterization of L-PBF lattice structures geometric defects. *Procedia CIRP 100* **2021**, *100*, 846–851. [CrossRef]
34. Soldo, M. A Newly Modeling Procedure for Different Morphological Distributions of Unit Cells in Lattice Structure with Constant Relative Density. *Int. J. Innov. Solut. Eng.* **2025**, *1*, 25. [CrossRef]
35. Inoma, A.; Ibhadode, O. A Novel Dual Curved Cubic (DCC) structure with improved compressive strength. *Int. J. Mech. Sci.* **2025**, *294*, 110186. [CrossRef]
36. Porterfield, A. Crystallon. Available online: https://www.food4rhino.com/en/app/crystallon?page=2&lang=de#commentsSection (accessed on 8 July 2025).
37. EcrLabs, Dendro. Available online: https://www.ecrlabs.com/dendro/ (accessed on 8 July 2025).
38. Kurtz, A. Intralattice. Available online: http://www.intralattice.com/ (accessed on 8 July 2025).
39. Pryor, M. Pufferfish. Available online: https://www.grasshopper3d.com/group/pufferfish (accessed on 8 July 2025).
40. Maconachie, T.; Leary, M.; Lozanovski, B.; Zhang, X.; Qian, M.; Faruque, O.; Brandt, M. SLM lattice structures: Properties, performance, applications and challenges. *Mater. Des.* **2019**, *183*, 108137. [CrossRef]
41. Jaroš, J.; Vaverka, O.; Senck, S.; Koutný, D. Influence of Process Energy on the Formation of Imperfections in Body-Centered Cubic Cells with Struts in the Vertical Orientation Produced by Laser Powder Bed Fusion from the Magnesium Alloy WE43. *Micromachines* **2024**, *15*, 278. [CrossRef] [PubMed]
42. Siebertz, K.; Van Bebber, D.; Hochkirchen, T. *Statistische Versuchsplanung*; Springer: Berlin/Heidelberg, Germany, 2017. [CrossRef]
43. Montgomery, D.C. *Design and Analysis of Experiments*; John Wiley & Sons, Inc.: Hoboken, NJ, USA, 2013; Volume 8.
44. EOS GmbH. MDS EOS PA2200. Available online: https://www.eos.info/polymer-solutions/polymer-materials/data-sheets/mds-pa-2200 (accessed on 4 July 2025).

45. Decker, T.; Kedziora, S. Optimizing the Thickness of Functionally Graded Lattice Structures for High-Performance Energy Absorption: A Case Study Based on a Bicycle Helmet. *Appl. Sci.* **2024**, *14*, 2788. [CrossRef]
46. Papazoglou, D.P.; NeidhardDoll, A.T.; Pinnell, M.F.; Erdahl, D.S.; Osborn, T.H. Compression and Tensile Testing of L-PBF Ti-6Al-4V Lattice Structures with Biomimetic Porosities and Strut Geometries for Orthopedic Implants. *Metals* **2024**, *14*, 232. [CrossRef]
47. *ISO 178:2019*; Plastics—Determination of Flexural Properties. Austrian Standards International: Vienna, Austria, 2025.
48. Autodesk Inc. Fusion 360. Available online: https://www.autodesk.com/products/fusion-360/overview?term=1-YEAR&tab=subscription. (accessed on 8 July 2025).
49. Mostafa, K.G.; Momesso, G.A.; Li, X.; Nobes, D.S.; Qureshi, A.J. Dual Graded Lattice Structures: Generation Framework and Mechanical Properties Characterization. *Polymers* **2021**, *13*, 1528. [CrossRef]
50. Yang, L.; Han, C.; Wu, H.; Hao, L.; Wei, Q.; Yan, C.; Shi, Y. Insights into unit cell size effect on mechanical responses and energy absorption capability of titanium graded porous structures manufactured by laser powder bed fusion. *J. Mech. Behav. Biomed. Mater.* **2020**, *109*, 103843. [CrossRef]
51. Li, Y.; Jiang, D.; Zhao, R.; Wang, X.; Wang, L.; Zhang, L.-C. High Mechanical Performance of Lattice Structures Fabricated by Additive Manufacturing. *Metals* **2024**, *14*, 1165. [CrossRef]
52. Di Prima, M.; Van Belleghem, S.; Badhe, Y.; Snodderly, K.; Porter, D.; Burchi, A.; Gilmour, L. Build parameter influence on strut thickness and mechanical performance in additively manufactured titanium lattice structures. *J. Mech. Behav. Biomed. Mater.* **2024**, *151*, 106369. [CrossRef]
53. Nechi, D. Evaluation of Different Software Solutions with Regard to the Generation of Lattice Structures for Additively Manufactured Components. Master's Thesis, University of Applied Sciences Vienna-Hochschule Campus Wien. Available online: https://resolver.obvsg.at/urn:nbn:at:at-fhcw:1-104773 (accessed on 22 July 2025).
54. Babacan, N.; Şeremet, H. Investigation of the load-bearing capacity of Co-Cr Lattice Structures fabricated by selective laser melting. *Int. J. 3D Print. Technol. Digit. Ind.* **2022**, *6*, 286. [CrossRef]
55. Distefano, F.; Pasta, S.; Epasto, G. Titanium Lattice Structures Produced via Additive Manufacturing for a Bone Scaffold: A Review. *J. Funct. Biomater.* **2023**, *14*, 125. [CrossRef] [PubMed]
56. Mantovani, S.; Giacalone, M.; Merulla, A.; Bassoli, E.; Defanti, S. Effective Mechanical Properties of AlSi7Mg Additively Manufactured Cubic Lattice Structures. 3D Print. *Addit. Manuf.* **2022**, *9*, 326–336. [CrossRef]
57. Noronha, J.; Leary, M.; Brandt, M.; Qian, M. AlSi10Mg hollow-strut lattice metamaterials by laser powder bed fusion. *Mater. Adv.* **2024**, *5*, 3751. [CrossRef]

applied sciences

MDPI

Article

An Intelligent Process Planning Method for Shaft Parts Based on Multi-Graph Fusion: From Feature Recognition to Process Route Generation

Zhenfeng Dai, Zhi Shou, Shuangling Ma and Xiaobo Peng *

Shenzhen Key Laboratory of Mold Advanced Manufacturing Technology, College of Mechatronics and Control Engineering, Shenzhen University, Shenzhen 518060, China; 2410095088@email.szu.edu.cn (S.M.)
* Correspondence: pengxb@szu.edu.cn

Abstract

Conventional process planning methods—typically driven by rules or expert heuristics—struggle to handle complex structural components, often yielding low efficiency and limited optimization. This paper proposes an intelligent process planning method for shaft parts based on multi-graph fusion. The framework integrates three core stages—machining feature recognition, machining scheme decision-making, and process route planning—into an end-to-end pipeline that transforms 3D models directly into executable machining routes. First, machining features are automatically identified using an attributed adjacency graph (AAG) representation coupled with a graph attention network (GAT). Next, the system leverages process knowledge and machining parameters to assign the optimal processing scheme for each feature. Finally, a sequence prediction model built on a machining-element directed graph (MEDGraph) generates process routes that satisfy manufacturing constraints. Experimental results demonstrate recognition and planning accuracies of 98.97% and 96.14%, respectively, underscoring the robustness and effectiveness of the proposed framework. This work establishes a unified pathway from design geometry to process execution, offering a powerful enabler for intelligent and adaptive manufacturing.

Keywords: machining feature recognition; process scheme decision-making; process route planning; graph neural networks

1. Introduction

Process planning plays a critical role in linking product design and manufacturing. For shaft parts, which are widely used in aerospace, automotive, and machine tool industries, the rationality of their machining process routes directly impacts product quality, manufacturing cost, and production efficiency. Despite continuous advancements in smart manufacturing technologies, process planning in manufacturing enterprises still relies heavily on manual experience, leading to issues such as low efficiency, weak reusability, and insufficient adaptability to complex manufacturing tasks [1]. Concurrently, the vast amounts of historical process data and standardized knowledge accumulated by enterprises have not yet been fully exploited.

With recent advances in artificial intelligence, research in process planning has gradually shifted from rule- and expert knowledge-based approaches towards data-driven methods. Technologies such as deep learning and graph neural networks have demonstrated promising performance in manufacturing feature recognition, process scheme recommendation, and process route optimization. However, current research still faces two major limitations. First, studies predominantly focus on isolated stages such as feature recognition, scheme decision-making, or operation generation, lacking an integrated modeling framework that spans the entire workflow. Moreover, the expression and utilization of manufacturing semantics (e.g., process attributes, tooling equipment, machining elements) are insufficient, making it difficult for planning outcomes to reflect knowledge associations within the engineering context.

To address the aforementioned issues, a multi-stage integrated process planning method for shaft parts is proposed in this study, which takes a 3D model as input and outputs a complete process route. The method encompasses three core steps: machining feature recognition, machining scheme decision-making, and process route generation, along with the construction of corresponding datasets and a model system.

The structure of this paper is as follows: Section 2 reviews related research. Section 3 introduces the fundamental concepts used in this study and details the dataset construction. Section 4 elaborates on the proposed methodology. Section 5 presents the experimental results and analysis. Section 6 provides the conclusions and future prospects.

2. Related Work

Intelligent process planning serves as a core link between product design (CAD) and manufacturing execution (CAM). As illustrated in Figure 1, it generally consists of two key stages: machining feature recognition (MFR) and process route planning. Machining feature recognition aims to extract geometrically meaningful features with manufacturing semantics from three-dimensional CAD models, thereby converting design information into an intermediate representation that can be interpreted by manufacturing knowledge. Based on the recognized features, process route planning determines machining sequences, methods, and parameters by considering process knowledge and resource constraints. Following this logical framework, this section reviews representative research on machining feature recognition and automated process route planning.

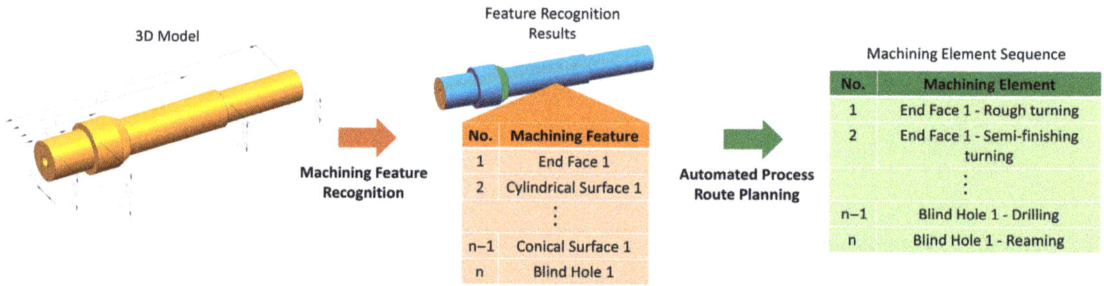

Figure 1. Flowchart of intelligent process route generation.

2.1. Machining Feature Recognition Methods

As illustrated in Figure 1, machining feature recognition represents the initial stage of intelligent process planning. Its core task is to automatically extract structural features with manufacturing semantics from three-dimensional models, thereby providing essential input for subsequent process planning. In recent years, with the rapid development of artificial intelligence technologies such as deep learning and graph neural networks (GNNs), research on MFR has shown a clear shift from traditional rule-driven approaches to data-driven methods, and from single-feature recognition to complex and interacting feature identification.

 https://doi.org/10.3390/app16020828

In terms of model representation, traditional point cloud and voxel structures are convenient for neural network processing but inherently suffer from the loss of boundary geometry and topological relationships. To preserve the integrity of manufacturing semantics, recent research has shifted toward constructing graph structures based on boundary representation (B-Rep). Lambourne et al. [2] proposed BRepNet, which directly takes the topological structure of B-Rep as input and employs coedge-based convolution kernels to achieve face-level semantic segmentation across 35,000 models, reaching an accuracy of 92.52%, significantly outperforming point cloud and mesh-based approaches under the same conditions. Ma et al. [3] introduced a multi-attribute face–edge graph (maFEG), encoding the full B-Rep as a graph structure and developing Sheet-metalNet to enable graph-level incremental learning. Similarly, Park et al. [4]'s DFGAT, Zhang et al. [5]'s BrepMFR, and Colligan et al. [6]'s Hierarchical CADNet are all based on B-Rep graph representations, further confirming its advantages in retaining geometric details and improving recognition accuracy.

In terms of recognition network architectures, GNNs have gradually replaced early convolutional networks and become the mainstream choice. Wang et al. [7] proposed the DeepFeature framework, which integrates rule-based methods with GNNs and employs graph convolution operations to learn local features of variable topological structures. Zheng et al. [8] combined Transformer self-attention mechanisms with GNNs to construct SFRGNN-DA, thereby improving recognition performance for complex geometries. Notably, Wu et al. [9] were the first to introduce a multi-task GNN, AAGNet, which takes geometric attributed adjacency graphs (gAAGs) as input and simultaneously performs semantic segmentation, instance segmentation, and base-face segmentation. On the MFInstSeg dataset containing more than ten thousand STEP files, AAGNet achieved state-of-the-art accuracy and effectively addressed the lack of instance-level segmentation capability in methods such as Hierarchical CADNet [6]. Building upon this, MFTReNet [10] further coupled semantic segmentation, instance grouping, and topological relationship prediction, significantly enhancing the modeling of complex topological associations among features. It should be noted that most existing machining feature recognition methods rely on ordinary graph structures to model pairwise relationships, which limits their ability to represent multi-feature co-occurrence and higher-order associations. Li et al. [11] proposed an adaptive multi-view hypergraph learning (AMH) framework, which effectively captures higher-order correlations through dynamic hyperedge construction and an attention-based fusion mechanism, thereby enhancing generalization performance across varying operating conditions. Although this method was originally developed for bearing fault diagnosis, its higher-order relationship modeling paradigm provides valuable insights for complex machining feature recognition.

Beyond model architecture, dataset quality and annotation richness also have a decisive impact on the performance of MFR systems. The MFTRCAD dataset released by Xia et al. [10] comprises over 20,000 STEP models and provides dual annotations at the instance level, including both feature segmentation and topological relationships. Yang et al. [12] constructed a volumetric feature dataset by converting feature recognition into a subgraph detection problem using Zone Graphs, thereby avoiding manually designed combinatorial rules. In addition, synthetic data generation techniques—such as the 24-class representative feature dataset in BrepMFR [5]—effectively alleviate the high cost associated with real data annotation.

Despite significant progress in geometric representation, network architecture, and task modeling for machining feature recognition in recent years, the overall research still faces several shortcomings. First, existing models predominantly rely on ordinary graph structures, which offer limited capability in characterizing complex geometric-topological

 https://doi.org/10.3390/app16020828

relationships, making it difficult to fully capture multi-feature interactions and higher-order association constraints. Second, there is a deficiency in cross-domain generalization ability, as the substantial distributional discrepancy between synthetic data and real-world parts significantly hinders the practical application of these models. Third, public datasets still exhibit limitations in terms of category coverage, topological annotation accuracy, and the scale of authentic data. To address the limitations of existing public datasets, this study constructs an extended dataset covering typical machining features of shaft-type parts. Each sample in this dataset is provided with both semantic segmentation labels and instance segmentation labels, thereby offering robust data support for model training and validation.

2.2. Automated Process Route Planning

After obtaining the machining feature recognition results, the process route planning stage (as shown in Figure 1) aims to generate the optimal machining sequence. This process requires balancing efficiency, quality, and cost under multiple constraints, including geometric, process, and resource limitations. With the development of intelligent manufacturing and industrial big data, traditional planning methods that rely on human expertise are gradually being replaced by data-driven and intelligent reasoning approaches.

Knowledge graphs explicitly represent manufacturing knowledge in a semantic form, serving as a core medium for reusable and interpretable process expertise. Liang et al. [13] proposed a knowledge-graph-based approach for modeling process design intent, capturing tacit experience through a "process design intent annotator" and supporting subsequent reuse. Xiao et al. [14] provided a systematic review of knowledge graph construction, refinement, validation, and generation techniques in CAPP, highlighting the emerging trend of integrating knowledge graphs with large language models. Zhang et al. [15] further combined knowledge graphs with reinforcement learning, transforming process decision-making into a sequence recommendation task on the graph, thereby improving the utilization efficiency of multi-source heterogeneous knowledge. Duan et al. [16] leveraged representation learning to extract typical machining process routines (TMPRs) from massive historical process routes, overcoming the limitations of traditional single-class classification based solely on part geometry and significantly enhancing process reuse.

GNNs have been widely applied to process planning due to their ability to simultaneously encode part topology, geometry, and Product Manufacturing Information (PMI). Zhang et al. [17] integrated stacked autoencoders with knowledge graphs to achieve end-to-end inference for macro-level process routes. Hussong et al. [18] proposed MaProNet, which employs a graph attention mechanism to directly learn from the attributed adjacency graph (AAG) of CAD models, thereby avoiding the information loss typically caused by voxelization. Wang et al. [19] constructed an attribute graph oriented toward machining features and used a graph convolutional network (GCN) to predict multiple feature-specific process routes in a single step, achieving an accuracy of 93.31% on an aerospace structural component dataset. Building on this, Zhang et al. [20] further combined convolutional graph neural networks with a deep reinforcement learning framework to enable efficient automatic process route generation for complex structural parts.

Reinforcement Learning (RL) has demonstrated notable advantages in rapid decision-making under dynamic resources and uncertain environments. Wu et al. [21] integrated operation sequencing and resource allocation into a unified decision matrix and employed an Actor–Critic framework to achieve fast responses to resource fluctuations in process planning. Guo et al. [22] proposed a multi-start exploration strategy that effectively alleviates the local optimum dilemma in large-scale problems. Xiao et al. [23] applied Proximal Policy Optimization (PPO) to learn sequencing patterns from historical process cases,

https://doi.org/10.3390/app16020828

and validated the generalization performance of the algorithm through experiments on 50 complex structural components. Furthermore, Zhang et al. [24] introduced Monte Carlo Tree Search for aerospace aluminum parts, enhancing state–action–reward modeling and thereby improving the exploration efficiency and solution quality of Deep Reinforcement Learning (DRL).

With the growing global emphasis on carbon emission reduction, machining time and energy consumption have emerged as critical objectives in process route optimization. Feng et al. [25] developed both cutting and non-cutting energy consumption models and proposed a staged integrated optimization strategy: first, Non-dominated Sorting Genetic Algorithm II (NSGA-II) was employed to simultaneously optimize cutting parameters to reduce energy consumption, shorten machining time, and ensure surface quality; subsequently, a genetic algorithm was applied to optimize operation sequencing, further decreasing non-cutting energy consumption. Chu et al. [26] introduced a prediction model based on radial basis function (RBF) networks and addressed the energy–time trade-off problem under multiple alternative process routes by combining Strength Pareto Evolutionary Algorithm 2 (SPEA2) with NSGA-II. Wang et al. [27], focusing on remanufacturing scenarios, proposed a machine spindle speed switching strategy and employed a reinforcement-learning-enhanced particle swarm optimization algorithm to simultaneously optimize process planning and scheduling, thereby achieving coordinated improvements in energy efficiency, economic benefits, and environmental performance.

Despite significant advancements in process route planning through knowledge graphs, graph neural networks, and reinforcement learning in recent years, several shortcomings persist. First, while existing knowledge graphs can explicitly represent process semantics, they remain inadequate in supporting dynamic evolution, cross-domain knowledge transfer, and deep integration with intelligent reasoning, thereby limiting their ability to fully facilitate complex process decision-making. Second, although research based on graph neural networks has enhanced the modeling of part geometry-process couplings, challenges remain in achieving a unified representation of multi-source heterogeneous information and efficient inference for large-scale parts. Third, while deep reinforcement learning demonstrates advantages in handling dynamic resources and uncertain environments, its convergence speed, global optimality, and interpretability still fall short of the requirements for industrial-level applications.

To address these gaps, the present study constructs a dedicated dataset incorporating process semantics and constraint information, thereby filling the void in existing research datasets. Methodologically, we design a decision-making model based on message-passing neural networks and a process route planning model that embeds process rules. This design enables unified representation of diverse process information while ensuring that process constraints are automatically satisfied during planning.

2.3. Process Planning Research for Shaft-Type Parts

Although significant progress has been made in general machining feature recognition and process planning methods, their performance remains highly dependent on specific application scenarios and part geometric characteristics. As typical rotational parts, shaft-type components have relatively standardized feature types. However, their machining sequences are subject to strict constraints and heavily rely on established process specifications and expert experience. Therefore, the core challenge for shaft-type parts lies not in feature diversity but in how to effectively embed domain knowledge to generate feasible and optimized machining sequences.

In recent years, research on process planning for shaft-type and rotational parts has developed several distinctive approaches. In terms of feature recognition, Zubair and

https://doi.org/10.3390/app16020828

Abu Mansor [28,29] employed a volumetric decomposition method, which decomposes the volume difference between the blank and the finished part into rough/finishing subvolumes, achieving systematic identification of internal and external features in both symmetrical and asymmetrical rotational parts. Elmesbahi et al. [30] proposed a hybrid graph-rule method based on the STEP-NC standard, enabling efficient recognition of complex interactive features such as axial and end-face features, thereby supporting seamless CAD/CAM integration.

Regarding process decision-making and knowledge representation, existing studies predominantly rely on knowledge bases and ontology modeling. Ma et al. [31] constructed an ontology-based knowledge base that explicitly defines causal relationships among features, operations, tools, and machine tools, and uses rule-based reasoning to automatically match manufacturing resources, enhancing system scalability. Li et al. [32] further integrated data-driven and knowledge-based reasoning by introducing a generative adversarial network to learn the distribution of process parameters, addressing the limitations of traditional case-based reasoning in adapting to new parts.

In terms of machining strategy innovation, Takamori et al. [33] proposed an automatic planning scheme combining turning and milling for organic-shaped parts. This approach prioritizes efficient rough machining via an optimal turning axis, supplemented by milling for finishing, offering an efficient machining paradigm for complex-profile shaft-type parts such as crankshafts.

In summary, while existing research has made progress in feature recognition, knowledge modeling, and specific strategies, it still faces two major limitations. First, there is an over-reliance on predefined rules, with insufficient capability for autonomously learning process sequences from data, resulting in limited flexibility. Second, most solutions remain isolated and have yet to form an end-to-end intelligent planning framework that integrates feature recognition, parameter optimization, and sequence generation. To address these limitations, the present study constructs a dedicated dataset for shaft-type parts and designs an intelligent planning model that uniformly represents multidimensional process information and automatically satisfies process constraints, thereby advancing the field toward a higher degree of automation.

3. Data Acquisition

Constructing large-scale and diverse datasets is essential for training and evaluating deep learning models. To meet the requirements of the proposed process route planning approach for shaft-type parts, three dedicated datasets were developed in this study: a machining feature recognition dataset, a feature–process decision dataset, and a process route planning dataset.

3.1. Construction of the Machining Feature Recognition Dataset

The creation of the machining feature recognition dataset primarily involves three stages: automatic generation of shaft-type parts, construction of the AAG, and definition of feature labels. The overall creation process is illustrated in Figure 2.

This feature recognition dataset is generated using parametric modeling technology. By programmatically constructing over 2000 three-dimensional models, it simulates the distribution patterns of machining features in real shaft-type parts. The models cover typical feature combinations, with both their geometric and topological structures validated for process rationality, thereby providing effective data support for deep model training. Although the sample size is approximately 2000, each part contains multiple feature nodes and topological relationships, such that the number of effective data units available for supervised learning far exceeds the nominal sample count. Combined with the advantages

https://doi.org/10.3390/app16020828

of graph neural networks in small-sample scenarios and the recognition accuracy exceeding 98% achieved in experiments, the scale of this dataset is sufficient to support adequate model convergence and generalization.

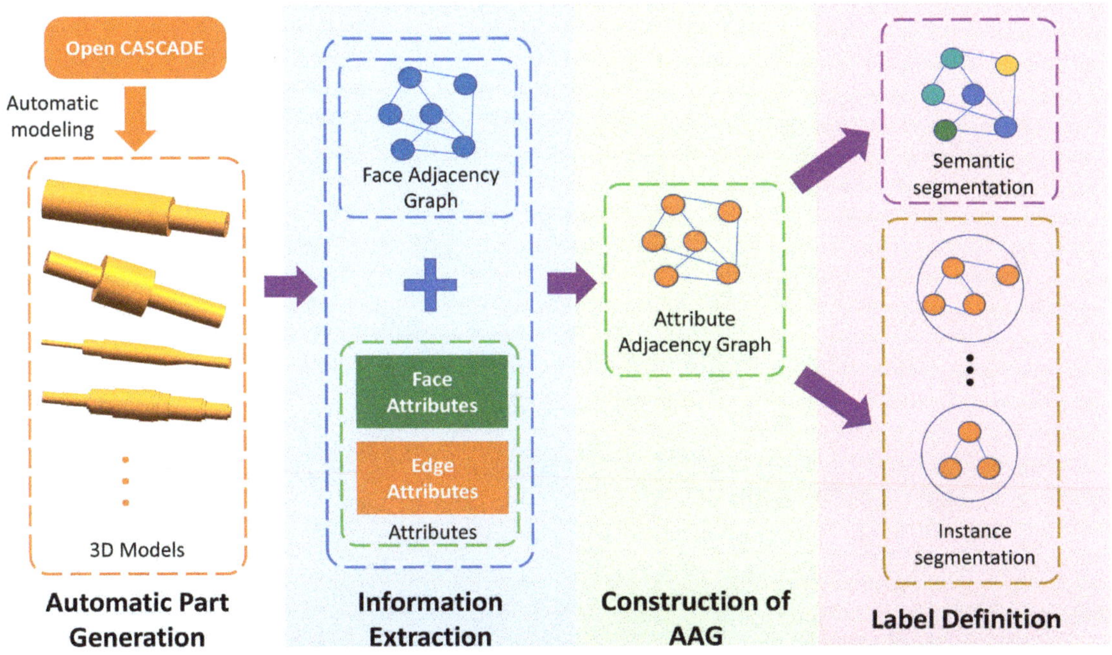

Figure 2. The workflow for building the machining feature recognition dataset.

The automatic generation of parts is designed to reflect actual turning machining logic while ensuring coverage of typical features across diverse geometric scales. Based on the practical sequence of turning operations, a cylindrical blank with an arbitrary length and diameter is first constructed. Subsequently, different diameters of external cylindrical features are generated by random segmentation. Following this, in accordance with the typical order of feature generation in turning operations, external conical features and relief groove features are randomly generated sequentially, and finally, through-holes or blind holes are generated. To ensure the rationality of the dataset, the generated shaft-type parts generally exhibit a spindle shape. At the same time, to guarantee full coverage of feature types and diversity in shape, structure, and scale, each three-dimensional model contains 4 to 20 machining features. Furthermore, the parameters of each feature (such as the diameter range of external cylinders, the taper range of external cones, and the size range of relief grooves) are randomly selected within reasonable intervals for actual shaft-type part machining, thus covering typical geometric feature combinations and diverse actual machining scales of shaft-type parts.

After generating the 3D models of the parts, this study adopts the method of AAG-Net [9] to structurally represent the 3D models using Attribute Adjacency Graphs (AAG). AAG integrates topological structure with face and edge attributes, enabling a complete description of machining features within the model. Specifically, an undirected graph (FAG) is first constructed based on face adjacency relationships. Then, attribute information of faces and edges is extended to form a complete AAG, with its attribute definitions provided in Table 1.

https://doi.org/10.3390/app16020828

Table 1. Face and edge attributes.

Entity	Attribute	Description
Face	Type	The geometric type of the face (plane, cylinder, etc.)
Face	Area	The area of the face
Face	Centroid	The centroid coordinates of the face
Face	Loops	The number of loops in the face
Edge	Type	The geometric type of the edge (line, circle, etc.)
Edge	Length	The length of the edge
Edge	Convexity	The convexity of the edge (concave, convex, or smooth)

To support supervised training, the dataset provides two types of labels for each model: semantic segmentation labels and instance segmentation labels. The semantic segmentation labels assign each face to a specific feature category (e.g., plane, cylindrical surface, etc.). The instance segmentation labels, in the form of an adjacency matrix, identify the face connectivity relationships within the same feature instance and distinguish face nodes belonging to different feature instances.

3.2. Construction of the Feature–Process Decision Dataset

To train and validate the feature–process decision model, a new dataset was further constructed on the basis of the feature recognition dataset. This dataset incorporates enriched process information and machining schemes, forming the feature–process decision dataset. The overall construction workflow is illustrated in Figure 3.

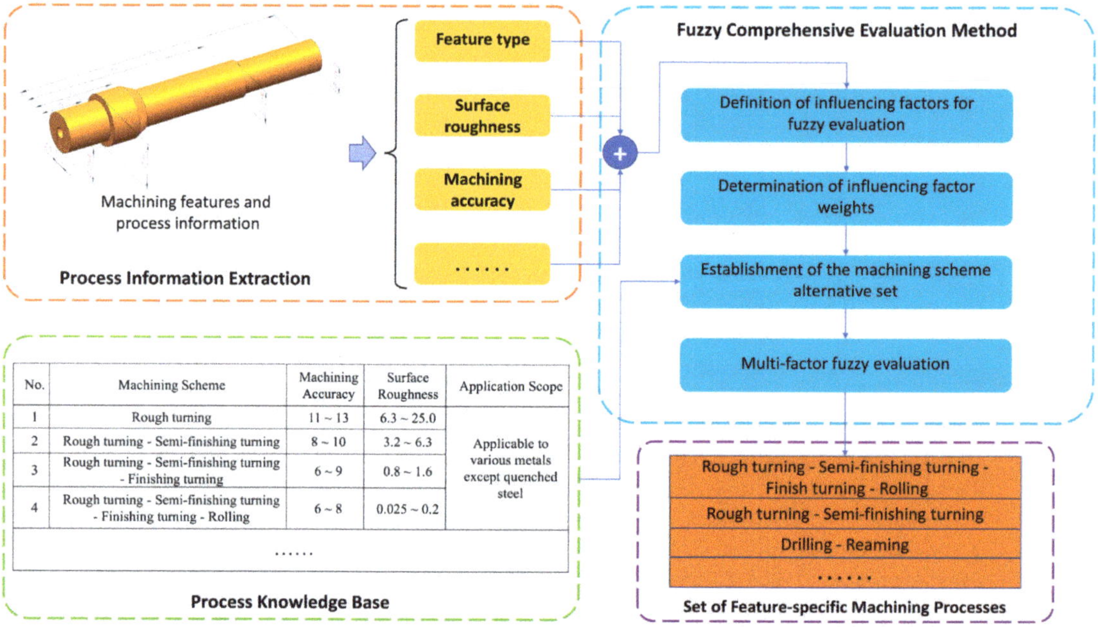

Figure 3. Workflow for constructing the feature–process decision dataset.

During the construction of the feature machining scheme decision dataset, a systematic feature machining scheme library was first established. This involved organizing common machining process paths and equipment parameters, clarifying the machining steps, applicable scopes, and typical process constraints for each scheme, thereby providing

https://doi.org/10.3390/app16020828

a knowledge base for subsequent scheme generation. Table 2 presents partial examples of external cylindrical surface machining schemes from this library. Subsequently, the existing machining feature recognition dataset was expanded. This expansion specifically included supplementing material information and related process parameters for the shaft-type parts. By integrating the process knowledge contained in the feature machining scheme library and applying a fuzzy comprehensive evaluation method, the optimal machining scheme was determined for each machining feature.

Table 2. Feature–process scheme library (partial).

No.	Machining Scheme	Machining Accuracy	Surface Roughness	Application Scope
1	Rough turning	11~13	6.3~25.0	Applicable to various metals except quenched steel
2	Rough turning–Semi-finishing turning	8~10	3.2~6.3	
3	Rough turning–Semi-finishing turning–Finishing turning	6~9	0.8~1.6	
4	Rough turning–Semi-finishing turning–Finishing turning–Rolling	6~8	0.025~0.2	
5	Rough turning–Semi-finishing turning–Grinding	6~8	0.4~0.8	Applicable to quenched steel and unquenched steel
6	Rough turning–Semi-finishing turning–Rough grinding–Precision grinding	5~7	0.1~0.4	

This study established an evaluation index system composed of material properties, feature types, machining accuracy, and surface roughness, which was applied to the fuzzy comprehensive evaluation of machining schemes. The weights assigned to each index are as follows: part material 0.444, feature type 0.312, machining accuracy 0.122, and surface roughness 0.122. Based on the determined indices and weights, a single-factor fuzzy evaluation was conducted to establish the fuzzy relationship matrix between machining schemes and feature requirements under each index. The weight vector and the fuzzy relationship matrix were then synthesized through computational operations to obtain the comprehensive evaluation results for each scheme. The optimal machining scheme was subsequently selected based on the ranking of these scores. Through the above process, the constructed dataset primarily consists of two parts: one is the extended attribute adjacency graph of the part and its feature attributes (including workpiece material, feature type, machining accuracy, surface roughness, etc.); the other is the labels, which represent the optimal machining scheme for each corresponding machining feature.

3.3. Construction of the Process Route Planning Dataset

To provide the data foundation required for building the process route planning model, a dataset was constructed that incorporates machining elements and their sequential dependencies. Based on feature–process schemes, the dataset was systematically decomposed into a collection of machining elements, with machining attributes and process constraints explicitly defined. Furthermore, machining workflows were represented in the form of graph structures, thereby supporting subsequent path generation and optimization. The overall workflow for constructing the process route planning dataset is illustrated in Figure 4.

Based on the previously determined feature–process schemes, each scheme was refined into multiple machining elements, thereby constructing a machining unit set that contains complete operation information. Each machining unit corresponds to a specific machining step and is associated with key attributes such as machining type, equipment and tool requirements, and machining stage. On this basis, typical process rules commonly applied in modern manufacturing practices—such as datum priority, roughing priority, surface

https://doi.org/10.3390/app16020828

finishing priority, and major-surface priority—were incorporated to further clarify the sequential dependencies among machining elements, thereby forming a structured set of process constraints.

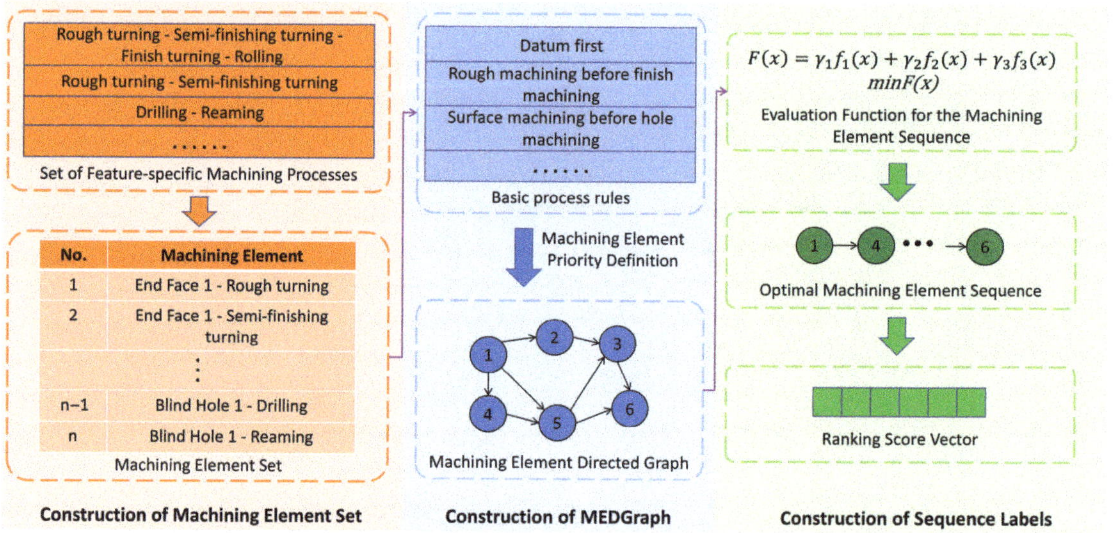

Figure 4. Workflow for constructing the process route planning dataset.

To formally represent machining elements and their dependencies, a machining-element directed graph (MEDGraph) was constructed. In this graph, each node represents a machining unit, while each directed edge indicates the precedence constraint between two machining elements. Each machining unit can be expressed as a quadruple:

$$me = (ID, f_i, S, T)$$

In this framework, ID denotes the unique identifier of a machining unit, f_i represents the machining feature, S denotes the machining stage (e.g., roughing, semi-finishing, and finishing), and T refers to the cutting tool. The construction of the MEDGraph involves initializing the graph, traversing machining elements, identifying constraint relationships, and creating the corresponding nodes and edges, thereby forming a complete directed graph structure.

To extract machining unit sequences from the MEDGraph, an improved topological sorting algorithm is employed, combined with a backtracking strategy to generate multiple candidate sequences. Subsequently, an evaluation function is constructed to select the optimal machining unit sequence from these candidates. The evaluation function considers machining cost (equipment, tools, time, etc.), machining efficiency (processing time, equipment utilization), and machining quality (accuracy, surface roughness) to compute an overall score for each sequence. The sequence with the highest score is selected as the optimal process route.

The dataset constructed based on the aforementioned method consists of two parts: first, the machining unit directed graph and its node features; second, the labels, which are the corresponding optimal machining unit sequences.

https://doi.org/10.3390/app16020828

4. Methods

To address the issues of heavy reliance on manual intervention and insufficient flexibility in traditional machining process route planning, this paper proposes an intelligent process planning method for shaft parts based on multi-graph fusion. The overall framework, as illustrated in Figure 5, comprises three core components: feature recognition, feature machining scheme decision-making, and machining process route planning.

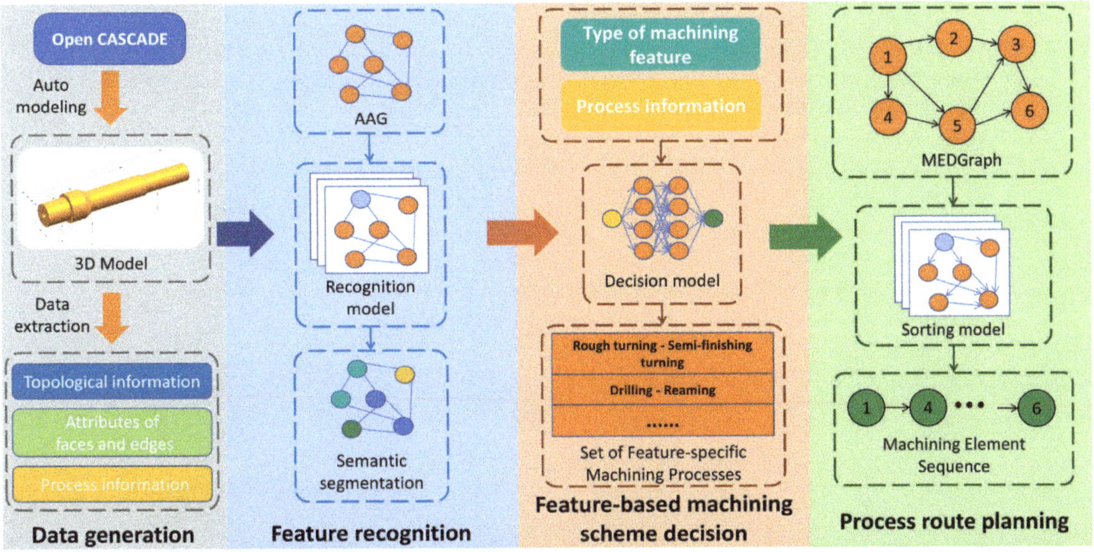

Figure 5. The pipeline of the proposed framework.

In the feature recognition stage, the system analyzes the 3D model of the shaft part to extract its topological and attribute information, constructing an AAG. With the aid of a graph neural network, machining features are classified and identified, thereby obtaining structured feature information.

In the process scheme decision-making stage, the method integrates the feature recognition results with the part's process requirements. It analyzes the machining constraints (such as material, accuracy, surface quality, etc.) for each feature and, based on the existing process scheme library, employs a decision-making model to match each feature with its optimal machining scheme. This results in a process scheme set covering all features.

In the process route planning stage, the system further decomposes the machining schemes into machining elements, constructing a dependency graph to clarify sequential constraints. A process planning model based on directed graphs is then utilized for learning and optimization, ultimately generating a machining unit sequence that satisfies the process requirements and is globally optimal, thereby completing the part's process route planning.

The proposed method achieves intelligent generation of a complete process route directly from a 3D model. It organically combines feature recognition, scheme decision-making, and route planning across three levels, providing a new approach for the efficient machining of shaft parts. The following sections will elaborate on these three aspects in detail.

4.1. Machining Feature Recognition

The machining feature recognition task involves understanding the structure of complex 3D geometric models, where different neighboring patches and topological relation-

https://doi.org/10.3390/app16020828

ships contribute unequally to feature semantics. For instance, critical datum planes or key connecting edges have a significantly greater impact on determining feature types than other auxiliary faces. Therefore, the model must be capable of automatically distinguishing the importance of different neighbors to more accurately extract local geometric patterns relevant to feature semantics. Based on these task characteristics, this paper selects the graph attention network (GAT). The attention mechanism of GAT can adaptively learn the influence of different neighbors on the target node, thereby highlighting key geometric elements when aggregating neighborhood information. Compared to traditional GCNs that use uniform neighborhood weights, GAT can effectively handle industrial CAD graph data with heterogeneous geometric structures and strong topological differences through learnable attention weights, and demonstrates improved expressive power and generalization performance even in scenarios with limited data.

Building on this, this paper proposes a GAT-based machining feature recognition model capable of processing nodes, edges, and their attributes in graph-structured data. Through two branches—semantic segmentation and instance segmentation—the model simultaneously achieves classification and localization identification of machining features. Figure 6 illustrates the overall architecture of the model, which consists of three core modules: the feature encoder, the graph encoder, and the feature classifier.

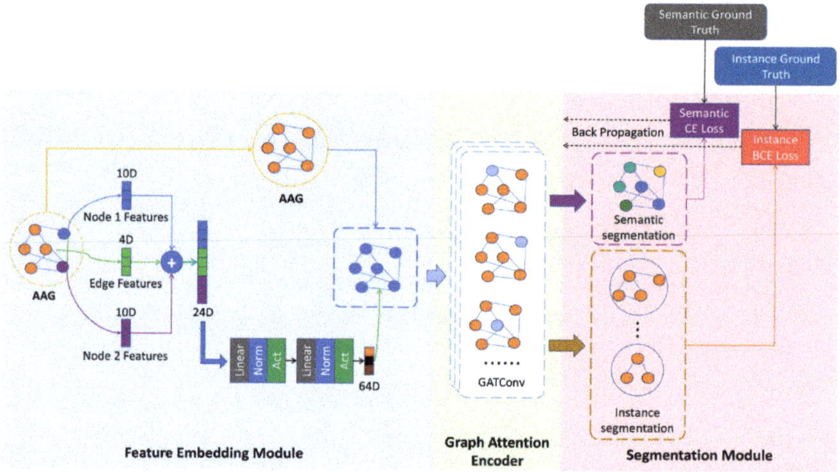

Figure 6. Architecture of the machining feature recognition model.

The feature encoder integrates local topological information of edges with their original attributes to generate enhanced edge representations. Specifically, for each edge, the source node features (10D), target node features (10D), and original edge features (4D) are concatenated to form a 24D composite vector. This vector is then transformed through a multilayer perceptron (MLP) to produce enhanced edge features in a latent feature space. By replacing the original edge features in the graph with these enhanced features, an enriched graph representation that captures local structural information is constructed. The enhanced edge features, together with the node features, serve as the input to the graph encoder, supporting subsequent message passing and feature aggregation.

The graph encoder employs a GAT architecture, which models dependencies within the graph by dynamically computing attention weights among nodes. In this architecture, each node adaptively attends to the importance of its neighbors, thereby effectively exploiting contextual information during feature updating. The attention mechanism in GAT mainly involves two steps: attention coefficient computation and feature aggregation. First,

 https://doi.org/10.3390/app16020828

the attention module evaluates the relevance between a target node and its neighboring nodes, assigning an importance weight to each neighbor. Then, the aggregation module performs a weighted summation of the neighbors' features according to these weights, yielding a summary vector that captures local context. This summary vector is typically combined with the node's original features (e.g., via concatenation or addition) and passed through a nonlinear activation function to complete the feature update. By stacking multiple graph attention layers, the encoder progressively captures more complex relational patterns among nodes, thereby achieving deep modeling of graph-structured features.

The feature classifier consists of two components: semantic segmentation and instance segmentation. For semantic segmentation, a Softmax classifier is employed to assign each node to its corresponding machining feature category (e.g., hole, slot, or cone). For instance, segmentation, an MLP is used to generate linkage scores between pairs of nodes, indicating whether two nodes belong to the same feature instance. The outputs of these two tasks are optimized using different loss functions and are jointly used to produce the final predictions of the model.

Regarding the loss functions, cross-entropy (CE) loss is adopted for semantic segmentation to quantify the discrepancy between predicted probabilities and ground-truth labels. The total semantic segmentation loss is computed as the average of the losses across all nodes and is denoted as L_s.

$$L_s = -\frac{1}{N}\sum_{i=1}^{N}\sum_{i=1}^{K} t_i log(p_i)$$

Here, K denotes the number of classes, which is 5 in the machining feature recognition dataset. p_i represents the predicted probability vector for a node belonging to class i, while t_i is the one-hot label vector, where $\hat{y}_i = 1$ if the label corresponds to class i, and $\hat{y}_i = 0$ otherwise. N denotes the total number of nodes.

For instance, segmentation, the loss function is the binary cross-entropy (BCE) loss, which measures the discrepancy between the predicted linkage scores S and the ground-truth adjacency matrix A. The ground-truth adjacency matrix is a binary matrix, where each element explicitly indicates whether a connection exists between two nodes. The total instance segmentation loss, L_i, is obtained by averaging the BCE losses across all node pairs.

$$\mathrm{BCE(t,p)} = -\mathrm{tlog(p)} - \mathrm{(1-t)log(1-p)}$$

$$L_i = \frac{1}{N^2}\sum_{i,j=1}^{N} \mathrm{BCE}\left(A_{ij}, S_{ij}\right)$$

Here, A_{ij} denotes the ground-truth adjacency value between nodes i and j, while S_{ij} represents the predicted linkage score between nodes i and j. N is the total number of nodes.

4.2. Feature–Process Decision

After completing machining feature recognition, this study further selects the most suitable machining scheme for each feature. Since the determination of machining schemes depends on process constraint relationships between a feature and its local neighborhood—such as relative dimensions, structural connectivity, and machining compatibility—a model capable of effectively modeling the local structural dependencies of "node-edge-node" is required. Based on these task characteristics, this paper selects the Message Passing Neural Network (MPNN) as the core framework. MPNN integrates node and edge attributes through an explicit message-passing mechanism, allowing it to directly represent local dependencies related to process semantics. Moreover, compared to models relying on

https://doi.org/10.3390/app16020828

attention mechanisms, MPNN offers more stable computation and lower memory overhead, making it more suitable for the industrial data in this study, which involves a large number of nodes and relatively regular graph structures.

Building on this, the study constructs the feature machining decision model shown in Figure 7, which automatically selects the optimal machining scheme for each feature from the extended attribute adjacency graph. The model consists of three key components: a feature encoder, a graph encoder, and a machining scheme decision module. The input to the model is an extended attribute adjacency graph containing node and edge attributes, and the output is the corresponding machining scheme category for each feature. Table 3 lists the face and edge attributes included in this extended attribute adjacency graph.

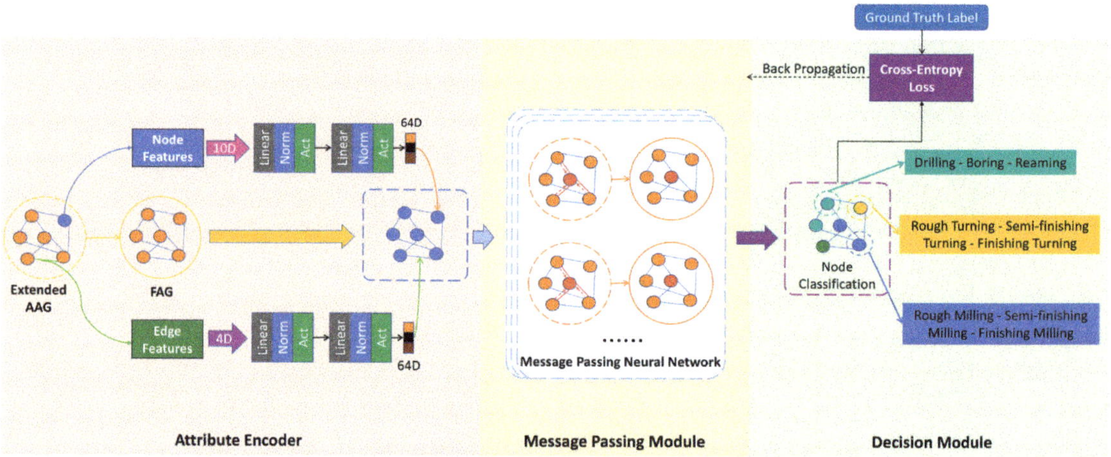

Figure 7. Architecture of the feature–process decision model.

Table 3. Face and edge attributes of the extended AAG.

Entity	Attribute	Description
Face	Type	The geometric type of the face (plane, cylinder, etc.)
Face	Area	The area of the face
Face	Centroid	The centroid coordinates of the face
Face	Loops	The number of loops in the face
Face	Material	The type of the workpiece material
Face	IT	The processing accuracy grade attribute of processing features
Face	Ra	The roughness attribute of machining features
Face	Feature	The type of the machining feature
Edge	Type	The geometric type of the edge (line, circle, etc.)
Edge	Length	The length of the edge
Edge	Convexity	The convexity of the edge (concave, convex, or smooth)

The attribute encoder employs an MLP to encode the node and edge attributes of the extended AAG, generating corresponding feature vectors. These encoded vectors are then assigned to the respective nodes and edges in the graph, forming an enhanced graph representation that serves as the input to the graph encoder.

 https://doi.org/10.3390/app16020828

The message passing module is built upon an MPNN architecture, which captures the dependencies among nodes and edges through iterative message passing. Each node updates its representation by aggregating information from its neighboring nodes. The message passing process involves three steps: message computation, aggregation, and update. Specifically, the message computation function generates messages based on node and edge features; the aggregation function collects messages from neighbors; and the update function refines the node representation using the aggregated messages. The graph encoder consists of multiple stacked message-passing layers, each employing different aggregation functions (e.g., sum, mean, or max) to capture diverse structural features of the graph.

In this study, the process scheme decision is formulated as a node classification task. The decision module applies a Softmax classifier to assign each node to one of the predefined machining scheme categories. The primary objective is to accurately classify all nodes into 15 machining scheme classes, thereby learning the complex mapping between graph-structured data and process schemes, and predicting the most suitable scheme for each machining feature.

Regarding the loss function, cross-entropy (CE) loss, denoted as L_{CE}, is also adopted for the process scheme decision.

$$L_{CE} = -\frac{1}{N}\sum_{i=1}^{N}\sum_{i=1}^{K} y_i^{(n)} log(p_i^{(n)})$$

Here, K denotes the number of machining scheme categories, which is 15 in the feature–process decision task. N is the number of training samples. $y_i^{(n)}$ represents the ground-truth label of the n-th sample, where $y_i^{(n)} = 1$ if the sample belongs to class i, and $y_i^{(n)} = 0$ otherwise. $p_i^{(n)}$ denotes the predicted probability that the n-th sample belongs to class i.

4.3. Process Route Planning

In the process route planning task, machining elements inherently form a directed graph structure with sequential dependencies. Therefore, a GCN that can directly process graph topology is selected as the core model. Compared to models such as GAT or MPNN, which rely on attention mechanisms or more complex message-passing methods, GCN demonstrates greater robustness in scenarios with limited industrial process data, clear machining logic, and stable structural relationships. It also incurs lower computational costs and can effectively capture local and global process dependency patterns through convolutional operations. Furthermore, the process constraint-aware convolutional mechanism proposed in this paper can naturally embed process semantics such as datum precedence and rough-finish sequencing within the GCN framework. This enables the model to simultaneously learn process rules while aggregating graph structural information, making it better aligned with the engineering characteristics of process route planning.

Consequently, this paper designs a process route planning model based on GCN, allowing it to accurately learn the mapping from "machining unit dependencies–process rules–optimal route" from historical process data. This enables the generation of machining routes that satisfy process logic constraints for new parts. The model takes the machining unit directed graph of the part as input and outputs the globally optimal machining unit sequence. The model architecture includes a feature encoding module, a graph convolutional module, and a ranking module, as illustrated in Figure 8.

The feature embedding module employs an MLP to encode the node attributes of the MEDGraph, generating corresponding feature vectors. These encoded vectors are then

 https://doi.org/10.3390/app16020828

assigned to their respective nodes, resulting in an enhanced MEDGraph that serves as the input to the graph convolution module.

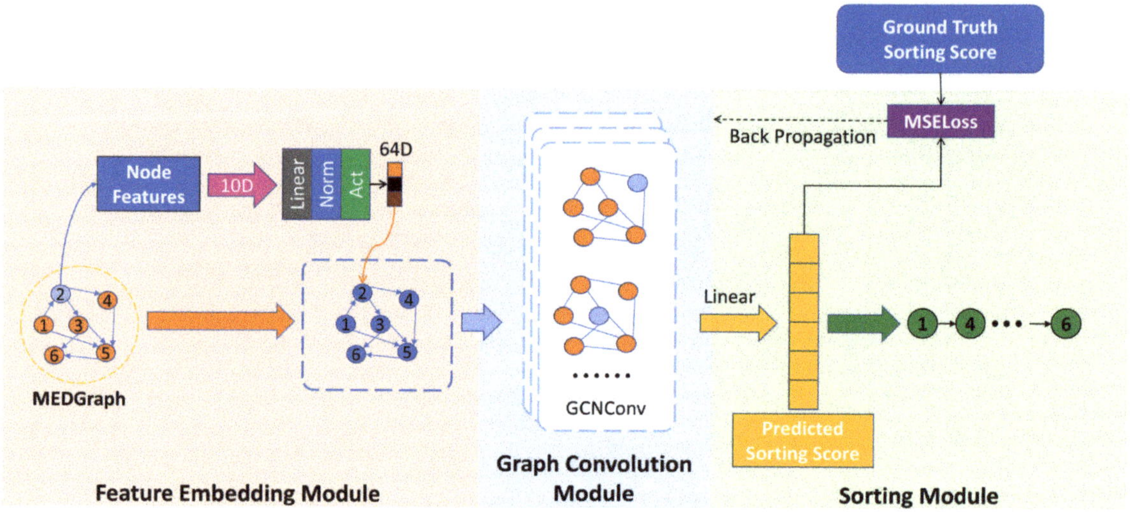

Figure 8. Architecture of the process route planning model.

The graph convolution module takes the enhanced MEDGraph as input and outputs updated node representations as well as their coordinate representations after convolutional operations. This module is composed of three primary convolutional layers. Batch normalization is applied to all convolutional layers, and a ReLU activation function is employed after each primary convolution layer. These layers are responsible for extracting shared feature representations from the input graph by propagating and aggregating information through node features and adjacency relationships, thereby updating the node representations.

In the graph convolution module, each convolutional layer enables nodes to aggregate information from neighboring nodes and connecting edges. As the network deepens, higher-level and more global structural features are progressively abstracted, enhancing the model's capacity to perceive the entire graph. This process allows each node to capture not only larger-scale local information but also global context, enriching the network's representational power. Typically, each convolutional layer is followed by a ReLU activation function to strengthen the non-linear expressive capability of the model. In addition, batch normalization is applied to all graph convolution layers to accelerate training and improve model stability.

The main responsibility of the sequencing module is to map the node feature vectors output from the graph convolution module into a ranking score vector. This ranking score vector represents the ranking scores of all nodes in the graph, with each node corresponding to a unique score. By sorting the nodes in ascending order of these scores, a reasonable machining sequence can be generated.

Specifically, the sequencing module uses a linear layer to transform the node feature vectors—obtained after multiple graph convolution operations—into ranking score vectors. During training, the model learns to accurately predict node sequences by minimizing

https://doi.org/10.3390/app16020828

the mean squared error (MSE) loss between the predicted ranking score vector and the ground-truth ranking score vector. The loss function is defined as follows:

$$\mathrm{MSE}(\mathbf{y}, \hat{\mathbf{y}}) = \frac{1}{N}\sum_{i=1}^{N}(y_i - \hat{y}_i)^2$$

Here, $\mathbf{y} = [y_1, y_2, \ldots, y_N]$ denotes the ground-truth ranking score vector, $\hat{\mathbf{y}} = [\hat{y}_1, \hat{y}_2, \ldots, \hat{y}_N]$ denotes the predicted ranking score vector, and N is the number of nodes in the machining unit directed graph. Based on the ranking score vector, the module can output the optimal sequence of machining elements, ultimately achieving the goal of process route planning.

5. Experiments and Results Analysis

5.1. Experimental Setup

To comprehensively validate the effectiveness of the proposed method, this study automatically constructs a dataset containing 3D models and complete process flows based on typical manufacturing process rules and historical machining documents. This comprehensive dataset consists of three sub-datasets designed to support the three tasks of feature recognition, machining scheme decision-making, and process route planning. These sub-datasets provide the necessary inputs and labels for training, validation, and testing at each stage of the model.

All datasets are randomly divided into a training set (80%, totaling 1600 samples), a validation set (15%, totaling 300 samples), and a test set (5%, totaling 100 samples). The training set is used for learning model parameters, the validation set for performance evaluation during training, and the test set for final generalization ability assessment. Furthermore, to enhance evaluation reliability, a ten-fold cross-validation method is applied to the training set. The training set is equally divided into 10 subsets, with 9 subsets used for training and 1 subset for validation in each iteration. This process is repeated 10 times, and the average performance is taken as the evaluation result of the model during the training phase. The experiments are conducted on a high-performance workstation equipped with an AMD Ryzen 7 7840HS processor, 1 TB of memory, and an NVIDIA RTX 4050 GPU. The deep learning platform used is PyTorch 2.1.0, the graph neural network tool is DGL 2.2.1, and the optimizer is ADAM.

5.2. Experimental Results

To verify the effectiveness of the sub-models in the tasks of feature recognition, process scheme decision, and process route planning, a comprehensive evaluation was conducted on the loss convergence, prediction accuracy, and generalization ability during training.

The loss and accuracy curves of the feature recognition model for semantic segmentation during training and validation are shown in Figure 9, while the corresponding confusion matrix on the test dataset is presented in Figure 10. The accuracy of all models was obtained through ten-fold cross-validation. In the feature recognition model, the semantic segmentation task achieved a training accuracy of 99.21% and a test accuracy of 98.97%. The instance segmentation task attained a training accuracy of 99.95% and a test accuracy of 99.71%. The training and validation performance of the instance segmentation task is illustrated in Figure 11. For the machining scheme decision-making task, the training and test accuracies were 98.58% and 98.14%, respectively. The corresponding loss and accuracy curves are shown in Figure 12, and the confusion matrix on the test dataset is given in Figure 13. The process route planning task achieved training and test accuracies of 96.87% and 96.14%, respectively. The loss and accuracy curves of the process route planning model during training and validation are presented in Figure 14. The accuracy for each task was calculated as follows: feature recognition was based on the proportion of correctly

https://doi.org/10.3390/app16020828

classified samples for semantic segmentation and the proportion of correctly segmented samples for instance segmentation; machining scheme decision-making was determined by comparing the consistency between predicted and actual schemes; and process route planning was evaluated based on the proportion of samples with exact matches.

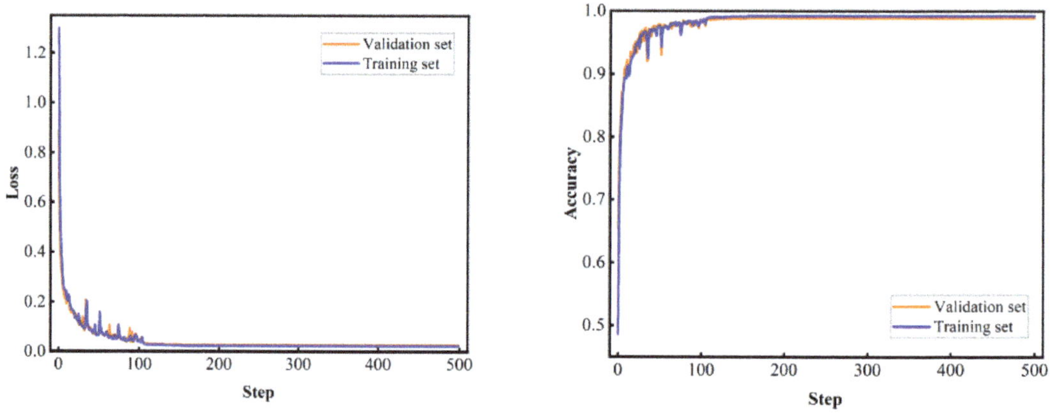

Figure 9. Loss and accuracy curves of the feature recognition model for semantic segmentation on the training and validation sets.

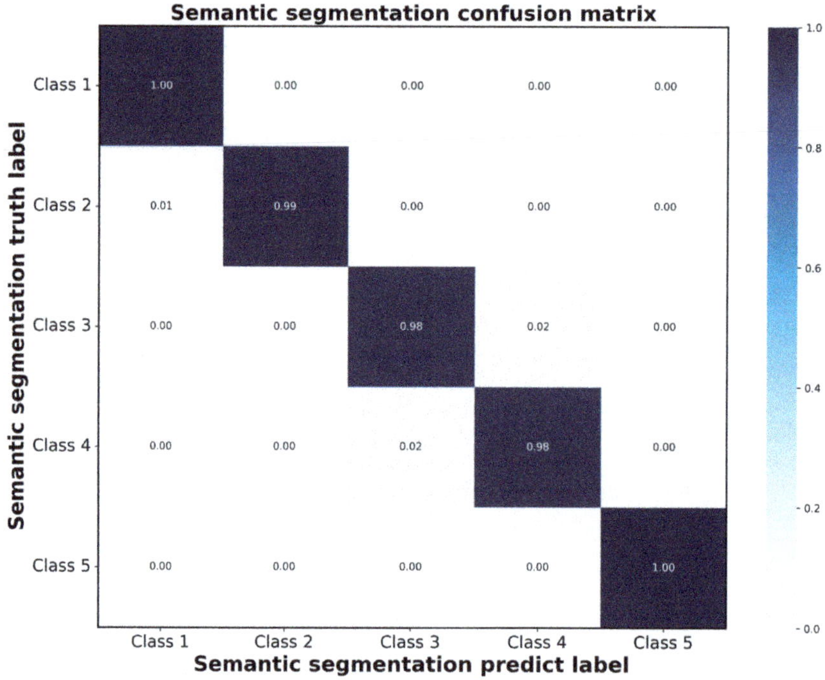

Figure 10. Confusion matrix for the semantic segmentation task on the test dataset.

The training process demonstrated that all sub-models converged well, with loss curves rapidly declining and then stabilizing. The training and validation accuracy curves remained closely aligned, showing no significant signs of overfitting. Furthermore, confu-

https://doi.org/10.3390/app16020828

sion matrices indicated that both the feature recognition and machining scheme decision-making models exhibited high classification precision and reliability on the test set.

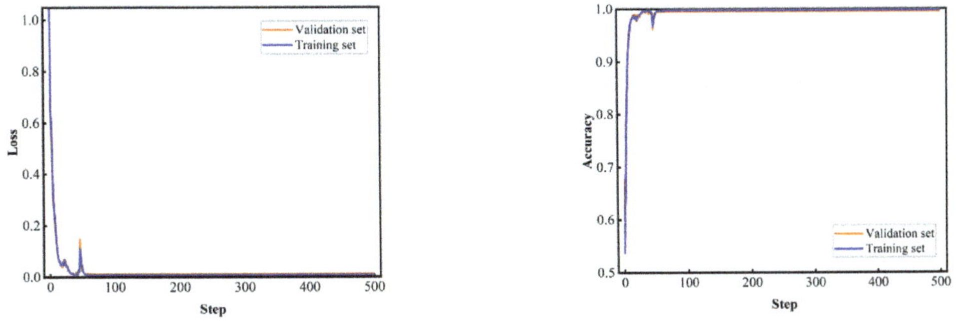

Figure 11. Loss and accuracy curves for instance segmentation of the feature recognition model on the training and validation sets.

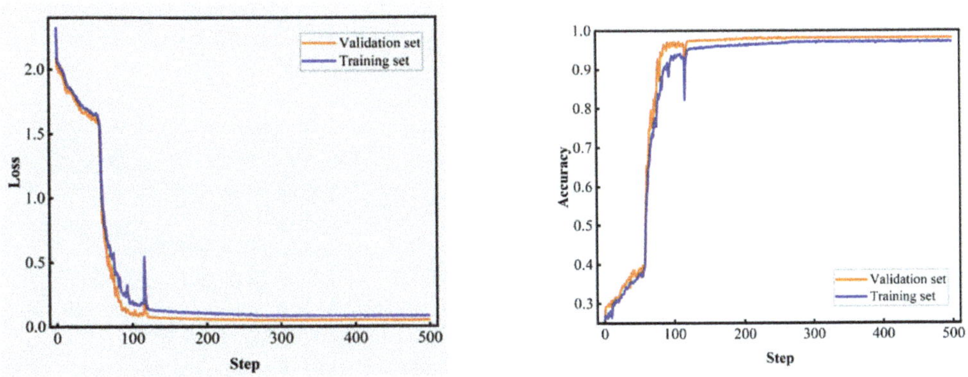

Figure 12. Loss and accuracy curves of the feature machining decision model on the training and validation sets.

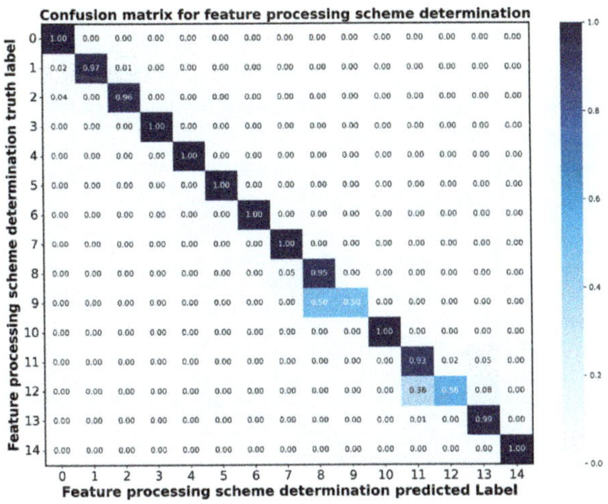

Figure 13. Confusion matrix for the machining scheme decision-making task on the test dataset.

https://doi.org/10.3390/app16020828

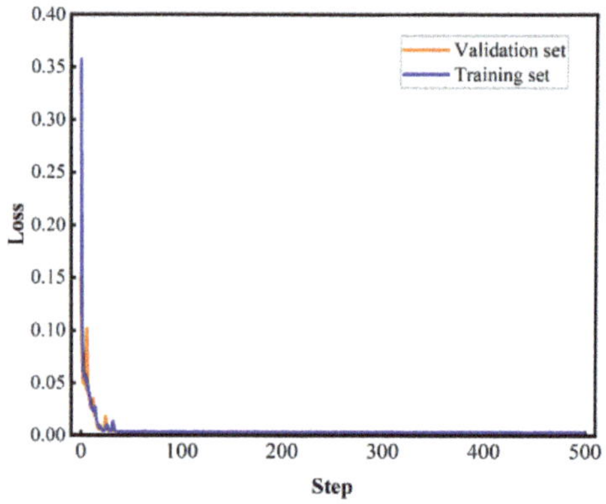

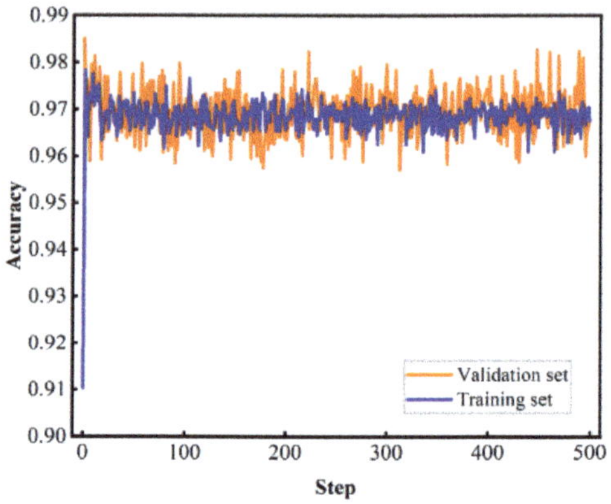

Figure 14. Loss and accuracy curves of the process route planning model on the training and validation sets.

In summary, all three sub-models achieved high accuracy and stable generalization performance, primarily attributed to their reasonable structural design and process semantic modeling. Feature recognition integrates geometric and topological information through attribute adjacency graphs and graph attention networks, enhancing the ability to identify complex structures. Machining scheme decision-making employs attribute encoding and nonlinear mapping to achieve precise matching under multiple constraints. Process route planning models operation dependencies based on the machining unit directed graph, balancing process rules with overall rationality. These designs collectively validate the effectiveness and feasibility of graph neural networks in integrated process planning.

When designing this process planning framework, we fully considered the requirements for computational efficiency and resource consumption in industrial applications. The relevant performance metrics are provided in Table 4. Based on the results in the table, it can be observed that the feature recognition module, built on the GAT architecture, maintains high inference efficiency despite involving attention weight calculations, particularly under the scale of attribute adjacency graphs typical for shaft parts. The machining scheme decision-making module effectively controls model size and memory usage by adopting the lightweight MPNN, making it more suitable for deployment in embedded process systems. The process route planning module utilizes the structurally concise GCN, whose local aggregation mechanism results in linear complexity, enabling stable inference times even on samples containing multiple machining elements. The entire framework, from 3D model input to process route generation, can be completed within the order of hundreds of milliseconds, meeting the real-time requirements of actual production. Furthermore, all three modules are constructed based on graph-structured data, endowing them with strong structural generalization and scalability. When new types of shaft parts are introduced, only the corresponding samples need to be supplemented, without requiring significant modifications to the overall architecture.

Table 4. Computational efficiency and resource consumption of each model.

Network	Number of Parameters	Average Inference Time Per Sample	Peak GPU Memory Usage
Feature recognition model	0.24 M	5.64 ms	0.01 GB
Feature machining decision model	0.19 M	5.32 ms	0.02 GB
Process route planning model	0.03 M	6.32 ms	0.02 GB

https://doi.org/10.3390/app16020828

To further validate the effectiveness of our approach, this study conducted comparative experiments on machining feature recognition using the publicly available MFCAD and MFCAD++ datasets, comparing the results with our self-constructed shaft part dataset. All experiments were based on the same model architecture and training configuration to ensure comparability of the results. Table 5 presents the recognition performance on the three datasets.

Table 5. Recognition performance on the three datasets under our proposed feature recognition model.

Dataset	Accuracy	F1-Score
The feature recognition dataset of this study	99.21%	98.80%
MFCAD	94.32%	93.37%
MFCAD++	90.87%	89.85%

The above results show that our method achieved the highest recognition accuracy of nearly 99% on the shaft part dataset. However, on public datasets such as MFCAD and MFCAD++, which feature more complex characteristics and are primarily focused on milling parts, the accuracy decreased compared to existing methods. This is mainly because the public datasets contain a wider variety of part types, with more complex feature morphologies and spatial relationships. In contrast, the design of our model is more tailored to shaft parts, utilizing lower input feature dimensions and a lighter network architecture. Consequently, it struggles to maintain the same level of performance in more complex scenarios. Overall, the comparative experiments validate the high reliability and accuracy of our method in the context of shaft parts, while also indicating that there is still room for improvement in the recognition of complex milling-type parts.

We also conducted various ablation studies to evaluate the impact of node attributes, edge attributes, and certain important attribute features on network performance. The ablation experimental results for each model are presented in Table 6.

Table 6. Ablation study results for each model.

Network	Attributes	Accuracy
Feature recognition model	full attributes	99.21%
	without node attributes	39.37%
	without edge attributes	98.23%
Feature machining decision model	full attributes	98.58%
	without machining feature type	83.41%
	without surface roughness and machining accuracy	59.18%
	without machining accuracy	60.67%
	without material type	65.57%
Process route planning model	full attributes	96.87%
	without node attributes	78.86%
	without edge attributes	95.91%
	without machining stage	96.52%
	without machining feature	96.43%

For the feature recognition model, the removal of node attributes led to a significant performance drop, indicating their critical role in distinguishing the geometric forms of shaft-type features. In contrast, the impact of edge attributes was relatively weaker. For the feature machining decision model, the absence of attributes such as machining accuracy, material type, and related process properties noticeably degraded model performance.

https://doi.org/10.3390/app16020828

This demonstrates that these attributes are directly linked to the logic of machining scheme selection and are core factors for the model's decision-making accuracy. For the process route planning model, node attributes also had a significant influence, as their removal led to a substantial decline in planning accuracy. Comparatively, removing individual node attributes like the machining stage or machining feature had a lesser effect on performance. This is primarily because the results of process route planning are determined by a combination of multiple feature attributes and process constraint relationships, exhibiting stronger multi-factor coupling characteristics. Overall, the ablation experiments validate the importance of various input attributes in different models. They also highlight that model design must adequately consider the contribution of each attribute to its specific task.

5.3. Case Study

To verify the effectiveness of the proposed intelligent process planning method for shaft parts based on multi-graph fusion, the typical shaft part shown in Figure 15 was selected as a case study for full-process analysis and experimental validation.

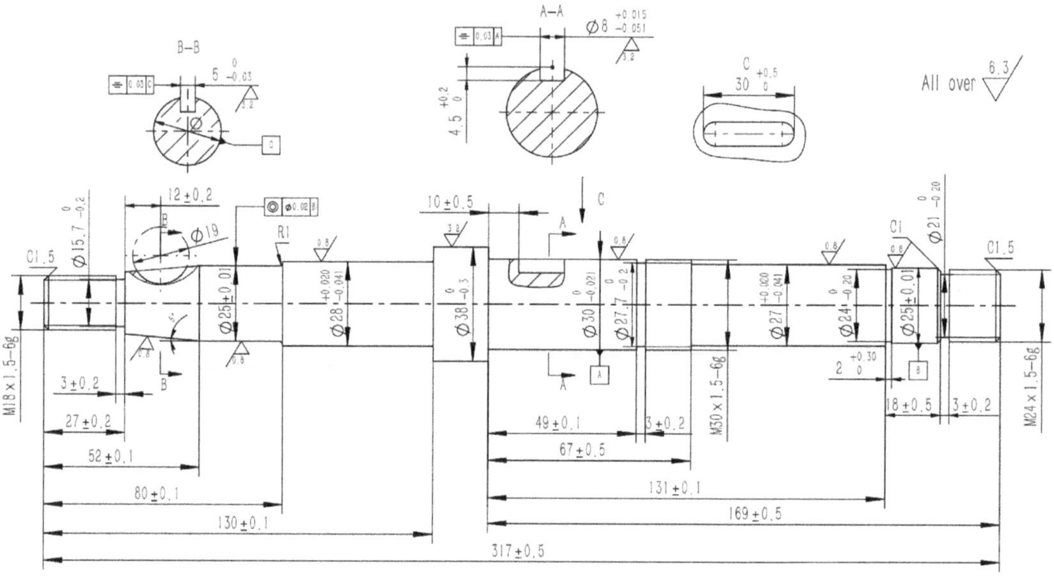

Figure 15. 3D model of the example part.

First, a 3D model is created based on the part engineering drawing, and corresponding process information is assigned to each feature. During the creation of the 3D model, features such as keyways, threads, chamfers, and fillets are removed. Subsequently, feature recognition is performed to identify all features within the part. The recognition results are shown in Figure 16. The model's recognition results are consistent with the actual features, indicating that the feature recognition model possesses strong capability in identifying machining features.

Next, by inputting the feature recognition results and machining process information into the feature machining decision model, the machining scheme for each machining feature can be obtained. The machining scheme set for the features of this part is presented in Table 7.

https://doi.org/10.3390/app16020828

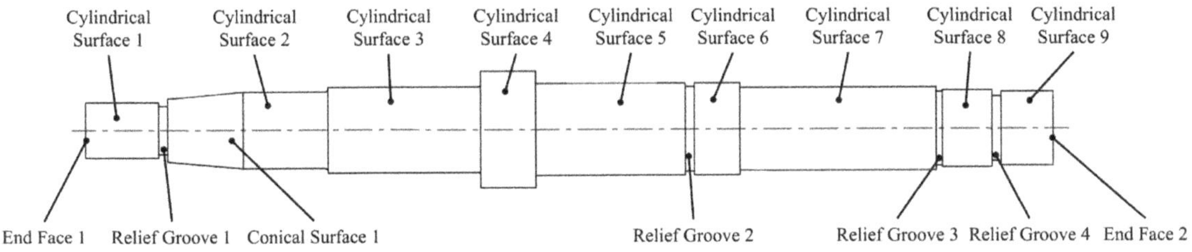

Figure 16. Machining feature recognition results for the example part.

Table 7. Machining feature recognition results of the example part.

Feature	Machining Scheme
End Face 1	Rough turning–Semi-finishing turning
End Face 2	Rough turning–Semi-finishing turning
Cylindrical Surface 1	Rough turning–Semi-finishing turning
Cylindrical Surface 2	Rough turning–Semi-finishing turning–Grinding
Cylindrical Surface 3	Rough turning–Semi-finishing turning–Grinding
Cylindrical Surface 4	Rough turning–Semi-finishing turning–Grinding
Cylindrical Surface 5	Rough turning–Semi-finishing turning
Cylindrical Surface 6	Rough turning–Semi-finishing turning
Cylindrical Surface 7	Rough turning–Semi-finishing turning–Grinding
Cylindrical Surface 8	Rough turning–Semi-finishing turning–Grinding
Cylindrical Surface 9	Rough turning–Semi-finishing turning
Conical Surface 1	Rough turning–Semi-finishing turning–Rough grinding–Finishing grinding
Relief Groove 1	Rough turning
Relief Groove 2	Rough turning
Relief Groove 3	Rough turning
Relief Groove 4	Rough turning

Based on the machining scheme set, a machining unit set is constructed and input into the process route planning model, thereby obtaining the predicted sequence of machining steps. Additionally, to meet the validation requirements of this study, the process documentation for this example part was organized and structured, resulting in an actual machining sequence derived from the experience of human process experts. A comparison between the model-predicted sequence and the actual machining sequence is presented in Table 8, from which it can be observed that differences occur mainly between the 14th and 27th machining steps, while the remaining steps are identical.

From the discrepancy analysis, it can be observed that after the "Cylindrical Surface 9—Rough turning" step, the model tends to continue the rough machining phase, thus sequentially scheduling steps such as rough turning of circular grooves and rough turning of cones. In contrast, in the actual process, operators prefer to first complete the semi-finishing of the external cylindrical surfaces by changing tools while maintaining the original clamping setup, before proceeding with operations like rough turning of grooves and cones.

https://doi.org/10.3390/app16020828

Table 8. Comparison between the predicted and actual machining sequences for the example part.

Predicted Machining Sequence	Actual Machining Sequence
End Face 1—Rough turning	End Face 1—Rough turning
End Face 2—Rough turning	End Face 2—Rough turning
End Face 1—Semi-finishing turning	End Face 1—Semi-finishing turning
End Face 2—Semi-finishing turning	End Face 2—Semi-finishing turning
Cylindrical Surface 3—Rough turning	Conical Surface 3—Rough turning
Cylindrical Surface 2—Rough turning	Cylindrical Surface 2—Rough turning
Cylindrical Surface 1—Rough turning	Cylindrical Surface 1—Rough turning
Cylindrical Surface 4—Rough turning	Cylindrical Surface 4—Rough turning
Cylindrical Surface 5—Rough turning	Cylindrical Surface 5—Rough turning
Cylindrical Surface 6—Rough turning	Cylindrical Surface 6—Rough turning
Cylindrical Surface 7—Rough turning	Cylindrical Surface 7—Rough turning
Cylindrical Surface 8—Rough turning	Cylindrical Surface 8—Rough turning
Cylindrical Surface 9—Rough turning	Cylindrical Surface 9—Rough turning
Relief Groove 1—Rough turning	Cylindrical Surface 4—Semi-finishing turning
Relief Groove 2—Rough turning	Cylindrical Surface 5—Semi-finishing turning
Relief Groove 3—Rough turning	Cylindrical Surface 6—Semi-finishing turning
Relief Groove 4—Rough turning	Cylindrical Surface 7—Semi-finishing turning
Conical Surface 1—Rough turning	Cylindrical Surface 8—Semi-finishing turning
Cylindrical Surface 4—Semi-finishing turning	Cylindrical Surface 9—Semi-finishing turning
Cylindrical Surface 5—Semi-finishing turning	Cylindrical Surface 3—Semi-finishing turning
Cylindrical Surface 6—Semi-finishing turning	Cylindrical Surface 2—Semi-finishing turning
Cylindrical Surface 7—Semi-finishing turning	Cylindrical Surface 1—Semi-finishing turning
Cylindrical Surface 8—Semi-finishing turning	Relief Groove 1—Rough turning
Cylindrical Surface 9—Semi-finishing turning	Relief Groove 2—Rough turning
Cylindrical Surface 3—Semi-finishing turning	Relief Groove 3—Rough turning
Cylindrical Surface 2—Semi-finishing turning	Relief Groove 4—Rough turning
Cylindrical Surface 1—Semi-finishing turning	Conical Surface 1—Rough turning
Conical Surface 1—Semi-finishing turning	Conical Surface 1—Semi-finishing turning
Cylindrical Surface 2—Grinding	Cylindrical Surface 2—Grinding
Cylindrical Surface 3—Grinding	Cylindrical Surface 3—Grinding
Cylindrical Surface 4—Grinding	Cylindrical Surface 4—Grinding
Cylindrical Surface 7—Grinding	Cylindrical Surface 7—Grinding
Cylindrical Surface 8—Grinding	Cylindrical Surface 8—Grinding
Conical Surface 1—Rough grinding	Conical Surface 1—Rough grinding
Conical Surface 1—Finishing grinding	Conical Surface 1—Finishing grinding

The case study demonstrates that the model-predicted machining sequence is highly consistent with the actual machining sequence in terms of overall order. Both encompass typical processes such as rough and semi-finishing machining and strictly adhere to the fundamental process principle of "roughing before finishing." This indicates that the proposed

 https://doi.org/10.3390/app16020828

process route planning model can not only effectively learn the topological dependencies between machining steps but also maintain process rationality and execution norms during sequence generation, thereby providing preliminary validation of the effectiveness of the proposed method.

6. Conclusions

This study proposes an intelligent process planning method for shaft parts based on multi-graph fusion, establishing a three-stage integrated technical system that progresses from machining feature recognition to feature machining scheme decision-making, and finally to process route planning. To support model training, a multi-source dataset covering feature recognition, scheme decision-making, and route planning was autonomously constructed, addressing the gaps in process semantic annotation and missing process parameters. Methodologically, a feature recognition model based on a GAT, a scheme decision-making model based on an MPNN, and a process-constraint-aware GCN embedding process rules were designed, achieving a unified representation of diverse process information and the integration of process constraints. Experimental results show that the proposed method can achieve fully automated conversion from a 3D model to a complete process route without manual intervention. It attains a high level of feature recognition accuracy, scheme decision-making rationality, and route planning consistency. The generated machining sequences show strong alignment with real process sequences, significantly enhancing the intelligence and industrial adaptability of process planning for shaft parts and providing effective technical support for the development of intelligent manufacturing.

Although the method proposed in this paper has achieved significant results in experimental validation, several issues remain for further research and refinement. First, it is necessary to further enhance the model's generalization ability for different types of parts and more complex, diverse machining scenarios, thereby improving its applicability in real industrial environments. Second, the interpretability and trustworthiness of the model require further strengthening to meet the practical demands for decision transparency and stability in industrial applications. Additionally, although the method has been validated on public datasets and real part case studies, comprehensive industrial-level manufacturing validation—such as systematic testing on actual machining equipment or within real production line environments—still poses certain challenges. This research is largely constrained by the accessibility of real process data and manufacturing resources. Future work will strengthen collaboration with industry to gradually introduce more real parts, actual process plans, and manufacturing process data, enabling a more comprehensive evaluation of the proposed method's performance in real production environments.

At the same time, for the tasks of machining scheme decision-making and process route planning, there is still a lack of structured, standardized, and publicly accessible benchmark datasets. This limitation hinders systematic comparison and reproducibility across different methods to some extent. Follow-up research will actively explore shareable data organization formats, progressively improve the transparency of synthetic data generation rules, and further validate and compare the proposed method based on available standard data. Ultimately, by integrating more efficient deep learning architectures, richer representations of geometric and process knowledge, and continuous industrial validation, we aim to build a more intelligent, reliable, and scalable automated process planning system.

Author Contributions: Conceptualization, X.P.; methodology, Z.D. and Z.S.; software, Z.D. and Z.S.; validation, Z.D. and Z.S.; investigation, S.M. and X.P.; manuscript writing and review, X.P., Z.D., Z.S. and S.M.; visualization, S.M. and Z.D.; project administration, X.P. All authors have read and agreed to the published version of the manuscript.

https://doi.org/10.3390/app16020828

Funding: This research was funded by the Shenzhen Natural Science Foundation (Basic Research Fund), grant number JCYJ20250604182633045.

Institutional Review Board Statement: Not applicable.

Informed Consent Statement: Not applicable.

Data Availability Statement: The datasets presented in this study are available from the author Z.D. via 2310295122@email.szu.edu.cn, on reasonable request.

Conflicts of Interest: The authors declare no conflicts of interest.

References

1. ElMaraghy, H.A. Flexible and reconfigurable manufacturing systems paradigms. *Int. J. Flex. Manuf. Syst.* **2005**, *17*, 261–276. [CrossRef]
2. Lambourne, J.; Willis, K.D.D.; Jayaraman, P.K.; Sanghi, A.; Meltzer, P.; Shayani, H. BRepNet: A topological message passing system for solid models. In Proceedings of the 2021 IEEE/CVF Conference on Computer Vision and Pattern Recognition (CVPR), Virtual, 19–25 June 2021; pp. 12768–12777. [CrossRef]
3. Ma, L.; Yang, J. Adaptive recognition of machining features in sheet metal parts based on a graph class-incremental learning strategy. *Sci. Rep.* **2024**, *14*, 10656. [CrossRef]
4. Park, J.H.; Lim, S.; Yeo, C.; Joung, Y.-K.; Mun, D. DFGAT for recognizing design features from a B-rep model for mechanical parts. *Robot. Comput.-Integr. Manuf.* **2025**, *93*, 102938. [CrossRef]
5. Zhang, S.; Guan, Z.; Jiang, H.; Wang, X.; Tan, P. BrepMFR: Enhancing machining feature recognition in B-rep models through deep learning and domain adaptation. *Comput. Aided Geom. Des.* **2024**, *111*, 102318. [CrossRef]
6. Colligan, A.R.; Robinson, T.T.; Nolan, D.C.; Hua, Y.; Cao, W. Hierarchical CADNet: Learning from B-Reps for Machining Feature Recognition. *Comput.-Aided Des.* **2022**, *147*, 103226. [CrossRef]
7. Wang, P.; Yang, W.-A.; You, Y. A hybrid learning framework for manufacturing feature recognition using graph neural networks. *J. Manuf. Process.* **2023**, *85*, 387–404. [CrossRef]
8. Zheng, X.; Chen, H.; He, F.; Liu, X. SFRGNN-DA: An enhanced graph neural network with domain adaptation for feature recognition in structural parts machining. *J. Manuf. Syst.* **2025**, *81*, 16–33. [CrossRef]
9. Wu, H.; Lei, R.; Peng, Y.; Gao, L. AAGNet: A graph neural network towards multi-task machining feature recognition. *Robot. Comput.-Integr. Manuf.* **2024**, *86*, 102661. [CrossRef]
10. Xia, M.; Zhao, X.; Hu, X. Machining feature and topological relationship recognition based on a multi-task graph neural network. *Adv. Eng. Inform.* **2024**, *62*, 102721. [CrossRef]
11. Li, Y.; Bwar, K.H.; Chai, R.; Tse, K.M.; Chai, B.X. Adaptive Multi-View Hypergraph Learning for Cross-Condition Bearing Fault Diagnosis. *Mach. Learn. Knowl. Extr.* **2025**, *7*, 147. [CrossRef]
12. Yang, D.; Li, Y.; Liu, X.; Deng, T.; Liu, C.; Xu, K. A data-driven approach for volume feature recognition based on cell graph. *Int. J. Comput. Integr. Manuf.* **2025**, *38*, 844–860. [CrossRef]
13. Liang, J.; Zhang, S.; Zhang, Y.; Huang, R.; Xu, C.; Wang, Z.; Zhang, H. A knowledge graph-based approach to modeling & representation for machining process design intent. *Adv. Eng. Inform.* **2024**, *62*, 102645. [CrossRef]
14. Xiao, Y.; Zheng, S.; Shi, J.; Du, X.; Hong, J. Knowledge graph-based manufacturing process planning: A state-of-the-art review. *J. Manuf. Syst.* **2023**, *70*, 417–435. [CrossRef]
15. Zhang, L.; Wu, H.; Chen, Y.; Wang, X.; Peng, Y. A knowledge-guided process planning approach with reinforcement learning. *J. Eng. Des.* **2024**, *36*, 1527–1550. [CrossRef]
16. Duan, J.; Duan, Y. Toward a framework of extracting typical machining process routines based on knowledge representation learning. *Adv. Eng. Inform.* **2024**, *60*, 102431. [CrossRef]
17. Zhang, Y.; Zhang, S.; Huang, R.; Huang, B.; Liang, J.; Zhang, H.; Wang, Z. Combining deep learning with knowledge graph for macro process planning. *Comput. Ind.* **2022**, *140*, 103668. [CrossRef]
18. Hussong, M.; Ruediger-Flore, P.; Klar, M.; Kloft, M.; Aurich, J.C. Selection of manufacturing processes using graph neural networks. *J. Manuf. Syst.* **2025**, *80*, 176–193. [CrossRef]
19. Wang, Z.; Zhang, S.; Zhang, H.; Zhang, Y.; Liang, J.; Huang, R.; Huang, B. Machining feature process route planning based on a graph convolutional neural network. *Adv. Eng. Inform.* **2024**, *59*, 102249. [CrossRef]
20. Zhang, H.; Wang, W.; Zhang, S.; Zhang, Y.; Zhou, J.; Wang, Z.; Huang, B.; Huang, R. A novel method based on deep reinforcement learning for machining process route planning. *Robot. Comput.-Integr. Manuf.* **2024**, *86*, 102688. [CrossRef]
21. Wu, W.; Huang, Z.; Zeng, J.; Fan, K. A fast decision-making method for process planning with dynamic machining resources via deep reinforcement learning. *J. Manuf. Syst.* **2021**, *58*, 392–411. [CrossRef]

 https://doi.org/10.3390/app16020828

22. Guo, K.; Liu, R.; Duan, G.; Liu, J. A Deep reinforcement learning method with multiple starting nodes for dynamic process planning decision making. *Comput. Ind. Eng.* **2024**, *194*, 110359. [CrossRef]
23. Xiao, B.; Zhao, Z.; Xu, B.; Li, Y.; Zhang, W.; Huan, H.; Su, H. A novel method for intelligent reasoning of machining step sequences based on deep reinforcement learning. *J. Manuf. Syst.* **2025**, *80*, 626–642. [CrossRef]
24. Zhang, H.; Wang, W.; Wang, Y.; Zhang, Y.; Zhou, J.; Huang, B.; Zhang, S. Employing deep reinforcement learning for machining process planning: An improved framework. *J. Manuf. Syst.* **2025**, *78*, 370–393. [CrossRef]
25. Feng, C.; Huang, Y.; Wu, Y.; Zhang, J. Feature-based optimization method integrating sequencing and cutting parameters for minimizing energy consumption of CNC machine tools. *Int. J. Adv. Manuf. Technol.* **2022**, *121*, 503–515. [CrossRef]
26. Chu, H.; Dong, K.; Yan, J.; Li, Z.; Liu, Z.; Cheng, Q.; Zhang, C. Flexible process planning based on predictive models for machining time and energy consumption. *Int. J. Adv. Manuf. Technol.* **2023**, *128*, 1763–1780. [CrossRef]
27. Wang, J.; Zheng, H.; Zhao, S.; Zhang, Q. Energy-aware remanufacturing process planning and scheduling problem using reinforcement learning-based particle swarm optimization algorithm. *J. Clean. Prod.* **2024**, *476*, 143771. [CrossRef]
28. Zubair, A.F.; Abu Mansor, M.S. Automatic feature recognition of regular features for symmetrical and non-symmetrical cylinder part using volume decomposition method. *Eng. Comput.* **2018**, *34*, 843–863. [CrossRef]
29. Zubair, A.F.; Abu Mansor, M.S. Embedding firefly algorithm in optimization of CAPP turning machining parameters for cutting tool selections. *Comput. Ind. Eng.* **2019**, *135*, 317–325. [CrossRef]
30. Elmesbahi, A.; Buj-Corral, I.; El Mesbahi, J.; Bensaid, O. New STEP-NC-compliant system to automate process planning for the turning process. *Int. J. Adv. Manuf. Technol.* **2023**, *128*, 2419–2457. [CrossRef]
31. Ma, H.; Zhou, X.; Liu, W.; Niu, Q.; Kong, C. A customizable process planning approach for rotational parts based on multi-level machining features and ontology. *Int. J. Adv. Manuf. Technol.* **2020**, *108*, 647–669. [CrossRef]
32. Li, Y.; Zhou, T. Research on Intelligent Planning Method for Turning Machining Process Based on Knowledge Base. *Machines* **2025**, *13*, 417. [CrossRef]
33. Takamori, R.; Nakatsuji, H.; Nishida, I. Automated process planning system to machine organic shapes by combining turning and milling. *J. Adv. Mech. Des. Syst. Manuf.* **2024**, *18*, JAMDSM0038. [CrossRef]

 https://doi.org/10.3390/app16020828

MDPI AG
Grosspeteranlage 5
4052 Basel
Switzerland
Tel.: +41 61 683 77 34

Applied Sciences Editorial Office
E-mail: applsci@mdpi.com
www.mdpi.com/journal/applsci

www.ingramcontent.com/pod-product-compliance
Lightning Source LLC
LaVergne TN
LVHW070451061126
851858LV00011B/213
* 9 7 8 3 7 2 5 8 8 3 5 5 4 *